Instructor's Guide to Accompany

MW01273682

BASIC TECHNICAL MATHEMATICS WITH CALCULUS
Fifth Edition

METRIC VERSION

Allyn J. Washington
Dutchess Community College

The Benjamin/Cummings Publishing Company, Inc.
Redwood City, California • Reading, Massachusetts • Fort Collins, Colorado
Don Mills, Ontario • Wokingham, U.K. • Amsterdam • Sydney
Singapore • Tokyo • Mexico City • Bogota • Santiago • San Juan

ISBN 0-8053-8893-1

ABCDEFGHIJK-MU-9543210

TABLE OF CONTENTS

PREFACE

This instructor's manual contains general comments and suggestions on the use of the material in the text and supplements, and the answers to all of the exercises. It is hoped that this information will aid instructors in their preparation and in designing their courses to fit the needs of the students.

The introductory comments and suggestions are related to the basic content and features of the text. Following is a discussion of the organization of the material, along with some suggestions as to some of the possible variations in material coverage to fit specific needs. Sections which may be omitted without loss of continuity are also mentioned. The supplements available for use along with the text are then listed.

The answers to the even-numbered exercises are given so that the instructor may give and readily check assignments for which the students do not have the answers available. The answers for the odd-numbered exercises are also included so that all answers are available in this manual for the convenience of the instructor. Also, some instructors may wish to have their students obtain this manual so that they may have the answers to all exercises available for checking their work.

Following the answers to the exercises is one supplementary topic of text material on Probability. It is from the third edition of Basic Technical Mathematics, and may be copied for use in any course in which a textbook published by Benjamin/Cummings Publishing Co. has been adopted and ordered from the publisher for the present semester or quarter.

At the end of this instructor's guide is a section on TECHDISK. Information and answers are given for the programs in this computer supplement to the text material.

It is hoped that the material presented here will aid the instructor, and thereby the student, in obtaining the maximum benefit from the text. I invite interested persons to share with me any suggestions they may have.

COMMENTS AND SUGGESTIONS

The material in Basic Technical Mathematics with Calculus, Metric Version, 5th ed., is generally sufficient for course work of up to three or four semesters, or four to six quarters. The only difference between this metric version and the regular version is that all units of measurement are SI metric units, or units which may be used with SI.

Chapters 1 through 19 provide a coverage of the basic topics in algebra and trigonometry. Chapter 20 covers the basic topics of analytic geometry, and Chapter 21 gives an introduction to statistics and curve fitting. Chapters 22 through 29 are devoted to a coverage of calculus. Certain selected supplementary topics and appendices are included after Chapter 29.

Each chapter is introduced by identifying the concepts which are to be developed in the chapter. Also, some of the important areas of technical applications are identified. A particular type of application is indicated in a photograph, and in Chapters 2 through 29 a problem related to this application is solved in an example later in the chapter.

Calculator material is integrated throughout the text. Specific operations are introduced where appropriate, and proper calculator use is shown in the examples. Also, Appendix D covers the basic operations of a scientific calculator. In addition, Appendix E gives selected computer programs in BASIC. These programs are keyed to the text material by use of a computer symbol, as shown at the right.

There is a special margin symbol (shown at the right) which is used to clearly indicate points with which students commonly make errors or which they tend to have more difficulty in handling. The text material which goes specifically with the symbol is printed in bold face italics. Although students vary widely in their abilities, they should be alerted to study these parts of the text with extra care.

NOTE ▷

Throughout the text, special explanatory comments in color have been used in the examples to emphasize and clarify certain important points. Arrows are often used to show clearly the part of the example to which reference is made. Also, certain important key terms and topics are noted in color in the margin for emphasis and reference.

The last section of each chapter consists of a summary of the important chapter equations, a complete set of review exercises for the chapter, and a practice test. The student may use the practice test to check his or her understanding of the material. Solutions to all problems on each test are given in the appendix after the answers to the odd-numbered exercises.

Specific mention is made of areas of application in many of the exercises and examples. Electricity, electronics, mechanics, and physics are noted more than other areas, since timing and application of knowledge of mathematics are generally more critical in these areas than in others. However, numerous other areas are noted, and it is recognized that many areas of technology could be included in connection with many of the applied exercises and examples. An index of applications for the exercises is given at the end of the book.

It should be recognized that some topics are important to the subsequent development of other mathematical skills with important technical applications. However, it is not necessary that the student (or the instructor) have a specific knowledge of the technical area from which any given problem is taken (with a few very specific exceptions where the necessary background is developed to a sufficient degree; e.g. Section 11-7 on ac circuits).

Organization of Topics

The text is organized such that certain topics - such as basic trigonometry, determinants, and complex numbers - are presented earlier than is normally the case. This makes it possible for students to take concurrent courses in which these and other topics can be used at appropriate times. It also eliminates the problem which arises when there is need for a given topic long before the material is taken up in the more traditional mathematics course.

Another feature of the organization of the text is that certain topics which could be introduced earlier in the text appear in the later chapters. Some of these topics are: systems of quadratics, equations with radicals, inequalities, variation, trigonometric identities, and inverse trigonometric functions. There are three primary reasons for including these in the later chapters. (1) The use and applications of these topics generally come later. (2) More elementary topics can be introduced in the course nearer the beginning, which is often very advantageous from the standpoint of their use in the allied technical courses. (3) Many of these topics tend to be somewhat troublesome to students, and they are taken up after the student has developed some of the basic mathematical techniques. This should aid in the understanding of these topics.

Many users of the earlier editions teach the material primarily as it is arranged in the text. However, other users have found that numerous variations of the order of topics may also be effectively followed. The topics included, and the order in which they are presented, can be selected to fit the needs of the course and the curriculum. Some of the possible variations of topic coverage and arrangement are as follows.

For courses which do not require an early introduction to trigonometry, Chapter 3 may be covered after Chapter 6. It is also possible to cover Chapters 3, 7, 8, and 9 after Chapter 10. However, at least Chapters 3 and 7 should be covered before Chapter 11 on complex numbers.

For courses which require a more complete early coverage of trigonometry, Chapters 3, 7, 8, and 9 may be covered following Chapters 1 and 2.

Chapter 17 on variation may be covered following Chapter 1, as variation is often taught at this point in algebra. However, since many users prefer it later, and others do not cover it, this topic is included later in the text.

For courses requiring an early coverage of higher-order determinants, Sections 15-1 and 15-2 or all of Chapter 15 can be taken up after Chapter 4.

The order and extent of coverage of the material in Chapters 13 through 21 depend primarily on the intent and scope of the course.

For courses which are designed primarily for precalculus coverage, Chapters 1, 2, 3, 5, and 6 may be omitted or covered as review material.

For users who require only an introduction to calculus, the course can end with Chapter 25, 26, or 27. The additonal chapters provide a more complete coverage.

The supplementary topics included after the regular text sections may be covered if they are deemed appropriate for the course. The material on units of measurement and approximate numbers should be referred to when appropriate.

Appendix C gives a reference for formulas and figures from geometry. It can be used in regular class sessions if a brief review of geometry is required.

Depending on individual needs there are some sections and chapters which may be omitted, or covered in less detail, without loss of continuity. These sections depend largely on the requirements and aims of the course, and the following list of sections serves only as an indication of those sections which might be considered to be in this category: 2-5, 4-5, 4-7, 6-4, 9-4, 9-6, 11-5, 11-6, 11-7, 12-7, 15-1 to 15-7, 16-4, 16-5, 18-4, 19-5, 20-5 to 20-10, 21-1 to 21-6, 23-6, 24-7, 25-4, 25-5, 25-6, 26-4, 26-7, 27-5 to 27-8, 28-5, 28-6, 29-3, 29-11, and 29-12.

SUPPLEMENTS

In addition to this Instructor's Guide there are other supplements which are available for use with the texts. All are available from The Benjamin/Cummings Publishing Co. These supplements are as follows:

TECHDISK by Seaver/Thomas is a computer supplement which gives students additional practice on many of the topics presented in the text. Information and answers for TECHDISK are given in the final section of this instructor's guide.

The Student's Solutions Manual by Willbanks/Zeigler contains fully worked solutions to every other odd-numbered exercise in the text.

A Computerized Test Bank includes tests with solutions (by Edmond) for each chapter. A printed copy is also available from the publisher.

Introduction to Geometry by Washington is a module which provides a more detailed coverage of basic geometry.

ANSWERS TO EVEN-NUMBERED EXERCISES

Exercises 1-1, p. 5

2. rational, real; imaginary   4. rational, real; irrational, real   6. 4; $\sqrt{2}$

8. $\frac{\pi}{2}$; $\frac{19}{4}$ 10. $7 > 5$ 12. $-4 < 0$ 14. $-\sqrt{2} > -9$ 16. $-0.6 < 0.2$ 18. $6, -\frac{4}{7}$

20. $-\frac{3}{8}$; $\frac{b}{y}$ 22.

24.

26. $-|-6|$, -4, $-\sqrt{10}$, $\frac{1}{5}$, 0.25, $|-\pi|$

28. (a) positive integer, (b) positive rational number, (c) positive integer
30. (a) between -1 and $+1$, (b) to the left of -2 or to the right of $+2$
32. g is constant, v and s are variables   34. $L = 100x - y$ 36. For $I < 4$ A, $V > 12$ V

Exercises 1-2, 1-3, p. 11

2. 4   4. 6   6. -3 8. 12   10. -27 12. -40 14. 3   16. -4 18. -72
20. 8   22. 0   24. undefined   26. 22   28. -58 30. 6   32. -8 34. 28
36. undefined   38. commutative law of addition
40. associative law of multiplication   42. distributive law
44. commutative law of multiplication   46. negative   48. no; $2 - 4 \neq 4 - 2$
50. $5 \times 7 = 7 \times 5$ lb, commutative law of mult.   52. $3(600 + 50)$, distributive law

Exercises 1-4, p. 13

2. 25.5   4. 57.5   6. 0.000 898 5   8. 3.652
10. 0.879 842 8 = 0.879 842 8, commutative law of multiplication
12. 0.628 293 1 = 0.628 293 1, associative law of multiplication   14. -102
16. 0.680   18. 18.8   20. -0.0330 22. (a) 2.083 333 3,   (b) 2.083 333 3
24. $\pi = 3.141\ 592\ 7$, $\frac{22}{7} = 3.142\ 857\ 1$ 26. 0.125 252 5   28. error   30. 377 cm
32. 153 L/day

Exercises 1-5, p. 19

2. y^9 4. $3k^6$ 6. x^5 8. $\frac{1}{s^3}$ 10. x^{24} 12. n^{21} 14. a^5x^5 16. $27a^6$

18. $\frac{x^7}{y^7}$ 20. $\frac{27}{n^9}$ 22. 1   24. 6   26. $\frac{1}{10^3}$ 28. t^5 30. $-y^{15}$ 32. c^{16} 34. 3

36. $\frac{2}{c^3}$ 38. $\frac{1}{x}$ 40. $\frac{1}{9t}$ 42. $\frac{m^3}{9n^8}$ 44. $\frac{y^{10}}{4b^4}$ 46. a^5x^4 48. $\frac{a}{b^2}$ 50. -3 52. -146

54. 48.0   56. 690   58. $\frac{g}{8\pi^3 f^3 CM}$ 60. 1.42 cm

Exercises 1-6, p. 23

2. 68 000 000   4. 0.000 096 1   6. 8.40   8. 0.1   10. 5.6×10^6 12. 7×10^{-1}

14. 1.09×10^0 16. 9.08×10^{-5} 18. 3×10^2 20. 5×10^{-12} 22. 84.0 Mm

24. 3.80 µg   26. 8.36×10^{-12} s   28. 3.25×10^{11} m   30. 1.02×10^3
32. 4.8730×10^{17} 34. 1.6×10^4 Pa   36. 0.000 000 000 004 5 s
38. 1 000 000 000 000 000 000 000 000 000°C   40. 7.9×10^{-6}%
42. 40 700 000 000 000 km   44. 2.5×10^5 46. 2.81×10^{-15} cm^3 48. 1.633×10^8 Pa·cm

Exercises 1-7, p. 26

 2. 9   4. −6   6. 15   8. −30   10. 2   12. −2   14. 19   16. 53   18. $4\sqrt{2}$
20. $5\sqrt{2}$ 22. $3\sqrt{6}$ 24. $4\sqrt{10}$ 26. $24\sqrt{3}$ 28. $\frac{8}{3}$ 30. 13   32. $4\sqrt{3}$ 34. 61.34
36. 0.250   38. 4.526 Ω   40. 2.07 cm   42. 15.00 m   44. no

Exercises 1-8, p. 31

 2. −t   4. −2c + d   6. 4x − 3y + z   8. $3xy^2 − 3x^2y^2$ 10. 8 − 4n + p   12. 3a − b
14. $−4\sqrt{x} + y$ 16. 6x + 3y   18. 7x − 5y   20. a + b   22. $−7a^2 − 5t + 4st$ 24. −2a + 2b
26. 3 − 3a   28. −2x − 12y   30. −10 + 6v   32. −x + 9y   34. 2x + 5   36. 3a − b − x
38. 2x − 12   40. $−3a^2 + 24$ 42. $i_1 + 4i_2 − 2$ 44. $3B − 2\alpha$

Exercises 1-9, p. 33

 2. $2x^3y^4$ 4. $−32c^2s^4$ 6. $54p^3q^7$ 8. $6m^7n^3$ 10. 2px − 2qx   12. $−6b^3 + 3b^2$
14. $2a^3bc^2 − 3a^4b^2c$ 16. $b^2x^4 − 2b^2x^3 + b^2x^2$ 18. $36c^3g + 8c^3 − 4c^2g^2$ 20. $18s^2t^3 − 24st^4$
22. $a^2 + 8a + 7$ 24. $8t^2 − 10st − 3s^2$ 26. $12x^2 − 13x + 3$ 28. $5p^2 + 38pq − 16q^2$
30. $6y^3 − 27y^2 + 4y − 18$ 32. $−2a^2b^4 − 7ab^2t + 30t^2$ 34. $2x^3 + x^2 − 13x − 15$
36. $5a^3 + 2a^2c − 8ac^2 + 3c^3$ 38. $−5y^2 − 15y + 90$ 40. $−ax^4 − 4ax^3 + 7ax^2 + 28ax$
42. $x^2 − 6x + 9$ 44. $4m^2 + 4m + 1$ 46. $b^2 − 4bx^2 + 4x^4$ 48. $3a^2 + 24a + 48$
50. $27x^3 − 27c^2x^2 + 9c^4x − c^6$ 52. $x^6 − 4x^5 − 4x^4 + 32x^3 − 16x^2 − 64x + 64$ 54. c − p + 2
56. $−x^3 + 39x^2 − 144x + 864$

Exercises 1-10, p. 37

 2. $−18b^6c$ 4. $\frac{3n^3}{m}$ 6. $4t^2$ 8. $\frac{4}{3b^3}$ 10. mn − 3n   12. −a − 2n
14. $ay^2 + x^2 − 1$ 16. $\frac{2ab}{3} − \frac{b}{3}$ 18. $1 + \frac{4}{xy} − 6y^2$ 20. $3b^2 − \frac{1}{a}$ 22. 3x + 1   24. 2x − 7
26. 3x + 2, R = 4   28. 2x − 4   30. $x^2 + 7x + 10$ 32. 2x − 1, R = 3x   34. $x^2 + x + 1$
36. $3a + 4b, R = 14b^2$ 38. $\frac{1}{R_2} + \frac{1}{R_1} + \frac{1}{6}$ 40. $s^2 + 2s + 6 + \frac{16s + 16}{s^2 − 2s − 2}$

Exercises 1-11, p. 40

 2. 5   4. −9   6. −8   8. 6   10. 3   12. −2   14. $−\frac{1}{9}$ 16. $\frac{3}{2}$ 18. $\frac{1}{2}$ 20. $−\frac{35}{3}$
22. 10   24. $\frac{4}{17}$ 26. 0.5   28. 1.74   30. Not true for any x, contradiction
32. True for all x, identity   34. 3.93 V   36. 0.46 m

Exercises 1-12, p. 43

 2. $−\frac{d}{c}$ 4. $\frac{a + 3}{b}$ 6. $−3n^2 − 2$ 8. 12p − 21   10. C − bx   12. $\frac{F}{DL}$ 14. $\frac{PV}{nR}$
16. $\frac{2Q − A − S}{2}$ 18. $−\frac{2mu}{e}$ 20. $\frac{P_1L + P_2L − FL}{P_1}$ 22. $\frac{A_1 − A}{A}$ 24. $\frac{3T_2 − T}{3}$

26. $\dfrac{p - p_a + dgy_1}{dg}$ 28. $\dfrac{t_c + t_m - t_a}{t_m}$ 30. $\dfrac{1 - hr_c + ahr_c}{h}$ 32. $\dfrac{TS_1 - Q - W}{T}$

34. 476 W   36. 0.200 m²

Exercises 1-13, p. 48

2. 185 L/min, 135 L/min   4. 3.2 N, 9.6 N, 12.8 N   6. 1330 mg, 670 mg   8. 480 000 L
10. 12, 17   12. 30 cm, 120 cm   14. 220 mm by 110 mm   16. 1160 km/h   18. 112 km
20. 3600 parts, 2500 parts   22. 180 g   24. 46.8 m³

Review Exercises for Chapter 1, p. 50

2. 2   4. −18   6. 16   8. -5   10. −5   12. 4   14. $27x^{12}y^3$ 16. $\dfrac{3p^3}{q^3}$ 18. $\dfrac{27}{b^6}$

20. $-7y^3$ 22. $2\sqrt{17}$ 24. $3\sqrt{5}$ 26. 9.228×10^{-13} 28. 70.70   30. $-3xy - 6y$

32. $x + 16b$ 34. $2x^2 - 7xy - 4y^2$ 36. $4x^2 + 12xy + 9y^2$ 38. $2ax - 4x^2$ 40. $3a - 2b$

42. $x^2 - 2b + 2y - 3z$ 44. $2x^3 - 9x^2 + 10x - 3$ 46. $-16s^3 + 24s^2t - 9st^2$

48. $6xy + 21rx - 12sx$ 50. $3st - 2s^2 + 1$ 52. $2x + \frac{1}{2}$, $R = -\frac{49}{2}$ 54. $x^2 - x + 4$

56. $4x^2 - 6x + 2$, $R = -3$ 58. $x^2 + 15x - 19$ 60. $3x + 4y$ 62. 2   64. −10

66. 8   68. $\frac{6}{5}$ 70. $\frac{2}{5}$ 72. −0.655   74. 3.4×10^8 km   76. 1.56×10^7 units

78. 1.5×10^{-7} m   80. 0.0072 m   82. $\dfrac{5 - 6b}{7}$ 84. $\dfrac{-29 - 3b}{10}$ 86. $\dfrac{DA}{KI^2}$ 88. $\dfrac{V - Ir}{I}$

90. $\dfrac{mu - mv}{v}$ 92. $\dfrac{f - 2B}{2B}$ 94. $\dfrac{2aZ^2 - 2ak}{Z^2}$ 96. $\dfrac{V - V_0 + 3aV_0T_1}{3aV_0}$ 98. $2.64 \times 10^3 \, \Omega$

100. 121 m   102. $2rV - aV - bV$ 104. $Ai + 2Ai^2 + Ai^3 - R - 2Ri - Ri^2$

106. 160 cm³, 80 cm³, 320 cm³   108. 0.8 ppm, 3.2 ppm   110. 75 m by 95 m   112. 18 min

114. 25 kg   116. 89 g

Exercises 2-1, p. 58

2. $A = \frac{1}{4}\pi d^2$ 4. $c = \pi d$ 6. $V = \frac{8}{3}\pi r^2$ 8. $p = 4s; s = \frac{p}{4}$ 10. 1, −19   12. 1.8, 15

14. $-5, -\frac{13}{4}$ 16. 6, 50.8   18. 15, 6a − 3   20. $-10t + 7$, $5t + 12$ 22. −3.356, 1.064

24. 1.0875, −2.1189   26. Multiply value of independent variable by 2, then subtract 6.

28. Subtract 5 times the value of the independent variable from 8 and add this value to the fifth power of the value of the independent variable.

30. $s = f(t)$, $f(t) = \sqrt{t + 2}$ 32. $R = f(R_c)$, $f(R_c) = \dfrac{10R_c}{10 + R_c}$ 34. −2.4 cm   36. $\dfrac{200(R+10)}{(110 + R)^2}$

Exercises 2-2, p. 62

2. Domain: all real numbers; range: all real numbers $g(u) \geq 3$
4. Domain: all real numbers $r \geq -4$; range: all real numbers $F(r) \geq 0$
6. Domain: all real numbers $x < 2$; range: all real numbers $f(x) > 0$
8. Domain: all real numbers; range: all real numbers $T(t) \geq -1$
10. All real numbers except $n = 3$ 12. All real numbers $x \geq 2$ except $x = 3$
14. 2, not defined   16. 0, 4   18. 2, $\frac{3}{4}$ 20. $-\frac{1}{2}, 5$ 22. $C = 10w + 1000$

24. $p = 100c - 300$ 26. $n = 0.5x + 70$ 28. $T = 0.04I - 480$ 30. $y = 1250 - 0.5x$

32. $A = d^2 + \frac{1}{4}\pi d^2$ 34. All real numbers greater than 0 and less than 2400

36. $C = 500$ for $\ell \leq 50$ m, $C = 5\ell + 250$ for $\ell > 50$ m

Exercises 2-3, p. 66

2. $(3, -2)$, $(-\frac{7}{2}, \frac{1}{2})$, $(0, -4)$

4.

6. Isos. rt. triangle   8. Parallelogram

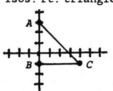
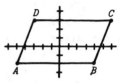

10. $\frac{9}{2}$ 12. $(4, -1)$ 14. on a line parallel to the x-axis, 3 units below

16. on a line parallel to the y-axis, 2 units to the left

18. on a line bisecting the second and fourth quadrants   20. 0

22. below the x-axis   24. above a line parallel to the x-axis, 4 units above

26. in second quadrant above a line one unit above x-axis   28. second, fourth

Exercises 2-4, p. 71

2.   4.   6.   8.   10.   12.

14.   16.   18.   20.   22.   24.

26.   28.   30.   32.   34.   36.

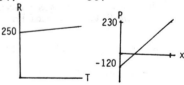

38.   40.   42.   44.   46.   48.

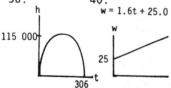

50. No   52. Yes

Exercises 2-5, p. 75

2.   4.   6. 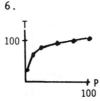 8.

10. (a) 139.4°C, (b) 3.7 min 12. (a) 9.7 V, (b) 16°C 14. 71°C 16. 3.7 cm
18. (a) 560 cm, (b) 2.8 m³/s 20. 260 cm 22. 0.45 24. 26 m² 26. 17.5 V

28. 3.9 m³/s

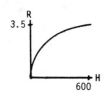

Exercises 2-6, p. 79

2. $\frac{4}{7}$ 4. $\frac{17}{3}$ 6. 0.7 8. 4.3 10. 0.0, 0.5 12. −1.2, 1.2 14. −0.4 16. −2.6
18. −1.2, 3.2 20. −1.6, 5.6 22. 2.1 24. −1.3, 0.0, 1.3 26. 1.9 28. −0.4, 2.4
30. after 7.5 years 32. 140 L 34. 18 s 36. 1.3 cm or 2.4 cm

Review Exercises for Chapter 2, p. 80

2. $A = 6e^2$ 4. $C = 50w + 600$ 6. $\frac{15}{2}$, 20 8. 43, 19 10. 8, $\frac{3v+1}{v+2}$ 12. $h^3 + 11h^2 + 36h$
14. −1778, −630.4 16. 0.0344, 0.0982
18. Domain: all real numbers except 0; range: all real numbers except 0
20. Domain: all real numbers $y \geq 0$; range: all real numbers $F(y) \leq 1$
22. 24. 26. 28. 30. 32.

      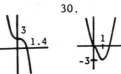

34. 2.0 36. −0.6, 8.6 38. no real zeros 40. 1.4 42. 0.0, 1.6 44. −5.0, 5.0
46. −3.7 48. 1.0, 2.0 50. 1.2 52. 1.0
54. (6,2) and (6,−3), or (−4,2) and (−4,−3) 56. 9.5%
58. 60. $N = 5000 + 7000t$ 62. 64. 66.

68. 3.46 m 70. $A = 21w - w^2$, 7 cm or 14 cm 72. $R_T = \frac{R_1^2 + 2.0R_1}{2R_1 + 2.0}$, 11 Ω

Exercises 3-1, p. 87

2. 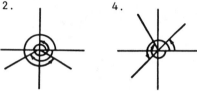   4.

6. 433°, −287°   8. 522°, −198°
10. 513°47′, −206°13′   12. 162.4°, −557.6°
14. 48.0°   16. 208.7°   18. 246.8°
20. 157.65°   22. −4.78°   24. 38.47°
26. 315°48′   28. −84°33′   30. 238°13′
32. 142°52′

34.   36.   38.   40.   42. 7.273°
44. 57°1′8″

Exercises 3-2, p. 91

2. $\sin\theta = \frac{12}{13}$, $\cos\theta = \frac{5}{13}$, $\tan\theta = \frac{12}{5}$, $\cot\theta = \frac{5}{12}$, $\sec\theta = \frac{13}{5}$, $\csc\theta = \frac{13}{12}$

4. $\sin\theta = \frac{7}{25}$, $\cos\theta = \frac{24}{25}$, $\tan\theta = \frac{7}{24}$, $\cot\theta = \frac{24}{7}$, $\sec\theta = \frac{25}{24}$, $\csc\theta = \frac{25}{7}$

6. $\sin\theta = \frac{15}{17}$, $\cos\theta = \frac{8}{17}$, $\tan\theta = \frac{15}{8}$, $\cot\theta = \frac{8}{15}$, $\sec\theta = \frac{17}{8}$, $\csc\theta = \frac{17}{15}$

8. $\sin\theta = \frac{2}{\sqrt{7}}$, $\cos\theta = \frac{\sqrt{3}}{\sqrt{7}}$, $\tan\theta = \frac{2}{\sqrt{3}}$, $\cot\theta = \frac{\sqrt{3}}{2}$, $\sec\theta = \frac{\sqrt{7}}{\sqrt{3}}$, $\csc\theta = \frac{\sqrt{7}}{2}$

10. $\sin\theta = \frac{5}{\sqrt{61}}$, $\cos\theta = \frac{6}{\sqrt{61}}$, $\tan\theta = \frac{5}{6}$, $\cot\theta = \frac{6}{5}$, $\sec\theta = \frac{\sqrt{61}}{6}$, $\csc\theta = \frac{\sqrt{61}}{5}$

12. $\sin\theta = \frac{1}{\sqrt{5}}$, $\cos\theta = \frac{2}{\sqrt{5}}$, $\tan\theta = \frac{1}{2}$, $\cot\theta = 2$, $\sec\theta = \frac{\sqrt{5}}{2}$, $\csc\theta = \sqrt{5}$

14. $\sin\theta = 0.808$, $\cos\theta = 0.589$, $\tan\theta = 1.37$, $\cot\theta = 0.729$, $\sec\theta = 1.70$, $\csc\theta = 1.24$

16. $\sin\theta = 0.5023$, $\cos\theta = 0.8647$, $\tan\theta = 0.5809$, $\cot\theta = 1.722$, $\sec\theta = 1.156$, $\csc\theta = 1.991$

18. $\frac{1}{2}\sqrt{3}$, 2   20. $\tan\theta = \frac{\sqrt{7}}{3}$, $\cos\theta = \frac{3}{4}$ 22. 0.945, 2.90   24. $\sec\theta = 1.81$, $\cot\theta = 0.663$

26. $\cos\theta = \frac{5}{13}$, $\cot\theta = \frac{5}{12}$ 28. $\csc\theta = \frac{1}{2}\sqrt{13}$, $\cos\theta = \frac{3}{\sqrt{13}}$ 30. $\cot\theta$ 32. $\frac{y}{r} \div \frac{x}{r} = \frac{y}{x} = \tan\theta$

Exercises 3-3, p. 95

2. $\sin 75° = 0.97$, $\cos 75° = 0.26$, $\tan 75° = 3.73$, $\cot 75° = 0.27$, $\sec 75° = 3.86$, $\csc 75° = 1.04$

4. $\sin 53° = 0.80$, $\cos 53° = 0.60$, $\tan 53° = 1.33$, $\cot 53° = 0.75$, $\sec 53° = 1.66$, $\csc 53° = 1.25$

6. 0.301   8. 0.588   10. 0.152 18   12. 0.68   14. 2.579   16. 1.12   18. 1.01
20. 1.029   22. 67.96°   24. 87.6°   26. 6.3°   28. 80.169°   30. 60.73°
32. 81.17°   34. 25.7°   36. 81.36°   38. 0.7401   40. 1.006   42. 0.731   44. 0.249
46. 67°   48. 57°   50. 0.660   52. 1.34   54. 6.2°   56. 51°10′   58. 83.4 cm
60. 32.9°

Exercises 3-4, p. 100

2.   4.

6. $a = 0.0283$, $B = 71.6°$, $b = 0.0851$

8. $A = 48.7°$, $B = 41.3°$, $b = 81.8$

10. $A = 25.7°$, $a = 0.314$, $c = 0.724$

12. $A = 55.2°$, $B = 34.8°$, $c = 7210$

14. $A = 77.40°$, $a = 17.98$, $b = 4.018$

16. $A = 51.67°$, $B = 38.33°$, $b = 7.833$ 18. $B = 38°$, $a = 11$, $c = 14$

20. $B = 38.64°$, $b = 295.2$, $c = 472.7$ 22. $A = 54.541°$, $B = 35.459°$, $a = 4.1430$

24. $A = 5.058°$, $b = 83,760$, $c = 84,087$ 26. $A = 28.5°$ 28. $x = 2774$

30. $B = 19°50'$, $b = 49.4$, $c = 146$ 32. $A = 72°50'$, $B = 17°10'$, $c = 7.17$

34. $\tan A = \frac{a}{b}$, $\tan B = \frac{b}{a}$, $c = \sqrt{a^2 + b^2}$ 36. $a = \sqrt{c^2 - b^2}$, $\cos A = \frac{b}{c}$, $\sin B = \frac{b}{c}$

Exercises 3-5, p. 103

2. 24.9 m   4. 0.281 m   6. 20.2°   8. 1570 m   10. 798 m   12. 51.3°
14. 1.96 m   16. 9.83 m   18. 47.1°   20. 3.07 cm   22. 462.6 m   24. 8.83 cm
26. 5.1 m   28. 7.95 m

Review Exercises for Chapter 3, p. 105

2. 608.3°, −111.7°   4. 352.4°, −367.6°   6. 174.75°   8. 321.45°

10. 65°24'   12. 126°15'

14. $\sin \theta = \frac{4}{\sqrt{41}}$, $\cos \theta = \frac{5}{\sqrt{41}}$, $\tan \theta = \frac{4}{5}$, $\cot \theta = \frac{5}{4}$, $\sec \theta = \frac{\sqrt{41}}{5}$, $\csc \theta = \frac{\sqrt{41}}{4}$

16. $\sin \theta = \frac{5}{13}$, $\cos \theta = \frac{12}{13}$, $\tan \theta = \frac{5}{12}$, $\cot \theta = \frac{12}{5}$, $\sec \theta = \frac{13}{12}$, $\csc \theta = \frac{13}{5}$

18. $\sin \theta = 0.927$, $\tan \theta = 2.47$ 20. $\sin \theta = 0.243$, $\sec \theta = 1.03$ 22. 0.763   24. 0.7557

26. 1.01   28. 5.416   30. 39.088°   32. 82.27°   34. 52.91°   36. 31.4°
38. $A = 21.9°$, $b = 2690$, $c = 2900$ 40. $A = 16.1°$, $B = 73.9°$, $b = 3.67$
42. $A = 74.3°$, $a = 12.1$, $b = 3.41$ 44. $A = 78.73°$, $B = 11.27°$, $c = 73.56$
46. $A = 85.62°$, $a = 74.18$, $c = 74.40$ 48. $A = 40.357°$, $B = 49.643°$, $c = 1118.7$

50. 9.36 m/s   52. 89.6 m²   54. 13.8 V·A   56. 14.2 m   58. 0.977 m²   60. 0.763 m
62. 1830 Ω   64. 4.71 min   66. 32°   68. 2.28 m, 5.96 m²   70. 30.8°   72. 58.2°
74. 14 m   76. 1.58 km

Exercises 4-1, p. 111

2. no; yes   4. yes; no   6. 2; $\frac{4}{7}$ 8. $-\frac{7}{2}$; −9   10. yes   12. yes   14. no

16. no   18. no   20. no

Exercises 4-2, p. 116

2. −6   4. 4   6. $-\frac{1}{2}$ 8. 0.2
10.   12.   14.   16.   18. $m = -4$, $b = 0$

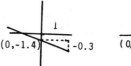

20. $m = \frac{4}{5}$, $b = 2$ 22. $m = 3$, $b = -\frac{7}{2}$ 24. $m = -\frac{1}{3}$, $b = \frac{1}{2}$ 26.   28.

      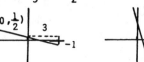

30.   32.   34. $C = -2d + 310$ 36.

   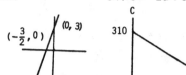   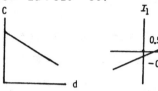

Exercises 4-3, p. 119

2. $x = 6.0$, $y = 2.0$ 4. $x = -4.0$, $y = -6.0$ 6. $x = 0.5$, $y = 3.3$ 8. $x = -1.5$, $y = 2.4$
10. $x = 5.0$, $y = 14.0$ 12. $x = 1.3$, $y = -0.8$ 14. $x = 4.0$, $y = -1.0$ 16. $x = 2.0$, $y = 3.0$
18. $x = 2.8$, $y = -0.6$ 20. $x = 1.7$, $y = 0.7$ 22. inconsistent   24. $x = 0.6$, $y = -0.4$
26. $x = 1.5$, $y = -0.3$ 28. dependent   30. $i_1 = 1.1$ A, $i_2 = 0.5$ A   32. 30 Mg, 10 Mg

Exercises 4-4, p. 125

2. $x = 5$, $y = 2$ 4. $x = -3$, $y = 4$ 6. $x = 2$, $y = -5$ 8. $x = 3$, $y = -\frac{5}{2}$ 10. $x = -1$, $y = \frac{2}{3}$
12. $x = -1.48$, $y = 0.104$ 14. $x = -2$, $y = 3$ 16. $x = 5$, $y = -3$ 18. $x = -1$, $y = -6$
20. dependent   22. $x = -\frac{7}{29}$, $y = -\frac{31}{29}$ 24. $x = 1.36$, $y = 0.272$ 26. $x = \frac{5}{3}$, $y = -\frac{1}{2}$
28. $x = -19$, $y = -16$ 30. inconsistent   32. $x = -0.067$, $y = -1.1$ 34. 165 cm, 65 cm
36. $c_1 = 10$ cm, $c_2 = 4$ cm   38. $F_1 = 40$ N, $F_2 = 20$ N   40. 12 700 m   42. 11 L, 9 L
44. No values can be found; system of equations is inconsistent; report is in error

Exercises 4-5, p. 131

2. -12 4. 5   6. 0   8. 162   10. 0.032   12. 121   14. $x = -2$, $y = 3$
16. $x = 5$, $y = -3$ 18. $x = -1$, $y = -6$ 20. dependent   22. $x = -\frac{7}{29}$, $y = -\frac{31}{29}$
24. $x = 1.36$, $y = 0.272$ 26. $x = \frac{5}{3}$, $y = -\frac{1}{2}$ 28. $x = -19$, $y = -16$ 30. inconsistent
32. $x = -0.067$, $y = -1.1$ 34. $A = 1.5T + 3.0$ 36. $x = 90$ MJ, $y = 60$ MJ
38. 3.27 m, 2.27 m   40. 6, 11   42. 348 m/s, 5200 m/s
44. $v_0 = 8.64$ m/s, $a = 3.06$ m/s²

Exercises 4-6, p. 137

2. $x = -1$, $y = 1$, $z = 3$ 4. $x = \frac{1}{2}$, $y = 2$, $z = -1$ 6. $r = \frac{5}{7}$, $s = \frac{6}{7}$, $t = 1$
8. $u = \frac{34}{3}$, $v = -\frac{31}{3}$, $w = -\frac{2}{3}$ 10. $x = \frac{1}{10}$, $y = \frac{1}{2}$, $z = -\frac{3}{10}$ 12. $x = \frac{1}{2}$, $y = \frac{5}{2}$, $z = -3$
14. $x = -2$, $y = -\frac{1}{4}$, $z = -\frac{3}{2}$ 16. $x = 1$, $y = 2$, $z = 3$, $t = 4$ 18. 3000 L/h, 5000 L/h, 6000 L/h
20. $-\frac{3}{22}$ A, $-\frac{39}{110}$ A, $\frac{27}{55}$ A   22. 2500 lines/min, 3250 lines/min, 3500 lines/min

24. system: $x + y = 500$, $z - y = 200$, $x + z = 700$ does not have unique solution

26. no solution   28. unlimited: $x = 1$, $y = -6$, $z = 0$

Exercises 4-7, p. 143

2. 84   4. 0   6. -232 8. -422 10. 26 660   12. -0.326 14. $x = 1$, $y = 2$, $z = -4$

16. $x = -1$, $y = 1$, $z = 3$ 18. $x = \frac{1}{2}$, $y = 2$, $z = -1$ 20. $r = \frac{5}{7}$, $s = \frac{6}{7}$, $t = 1$

22. $u = \frac{34}{3}$, $v = -\frac{31}{3}$, $w = -\frac{2}{3}$ 24. $x = \frac{1}{10}$, $y = \frac{1}{2}$, $z = -\frac{3}{10}$ 26. $x = \frac{1}{2}$, $y = \frac{5}{2}$, $z = -3$

28. $x = -0.253$, $y = -0.599$, $z = -0.278$ 30. $I_A = 0.71$ A, $I_B = -0.39$ A, $I_C = -0.32$ A

32. 100 mL, 150 mL, 250 mL

Review Exercises for Chapter 4, p. 145

2. -24 4. -3.674 6. -3 8. $-\frac{4}{7}$ 10. $m = \frac{2}{3}$, $b = -3$ 12. $m = 1$, $b = -\frac{8}{3}$

14. $x = 1.8$, $y = -2.4$ 16. $x = 2.4$, $y = -1.1$

18. $x = 2.2$, $y = 1.3$ 20. $x = 1.1$, $y = -0.2$

22. $x = 3$, $y = -1$ 24. $x = -4$, $y = \frac{2}{3}$

26. $x = \frac{17}{24}$, $y = -\frac{23}{48}$ 28. $x = \frac{60}{17}$, $y = -\frac{11}{17}$

30. $x = \frac{97}{58}$, $y = -\frac{7}{58}$ 32. $x = 49.0$, $y = 34.5$

34. $x = 3$, $y = -1$ 36. $x = -4$, $y = \frac{2}{3}$ 38. $x = \frac{17}{24}$, $y = -\frac{23}{48}$ 40. $x = \frac{60}{17}$, $y = -\frac{11}{17}$

42. $x = \frac{97}{58}$, $y = -\frac{7}{58}$ 44. $x = 49.0$, $y = 34.5$ 46. -105 48. 101 260

50. $x = 3$, $y = \frac{1}{2}$, $z = -2$ 52. $x = -\frac{11}{17}$, $y = \frac{7}{34}$, $z = -\frac{49}{17}$ 54. $u = -\frac{1}{2}$, $v = -2$, $w = -3$

56. $t = \frac{1}{3}$, $u = -3$, $v = \frac{4}{3}$ 58. $x = 3$, $y = \frac{1}{2}$, $z = -2$ 60. $x = -\frac{11}{17}$, $y = \frac{7}{34}$, $z = -\frac{49}{17}$

62. $u = -\frac{1}{2}$, $v = -2$, $w = -3$ 64. $t = \frac{1}{3}$, $u = -3$, $v = \frac{4}{3}$ 66. $x = -\frac{1}{2}$, $y = \frac{1}{5}$

68. $x = \frac{8}{9}$, $y = -\frac{16}{5}$ 70. 8   72. -4 74. $i_1 = 0.003\ 79$ A, $i_2 = -0.572$ A, $i_3 = 0.569$ A

76. \$1500, 8 %   78. $v = 0.6067T + 331.4$ 80. 5.44 m by 3.64 m   82. 6.7 L, 3.3 L

84. 0.013 g, 1.342 g, 0.395 g

Exercises 5-1, p. 152

2. $2ax - 6x$ 4. $6a^3 + 21a^2$ 6. $s^2 - 4t^2$ 8. $a^2b^2 - c^2$ 10. $49s^2 - 4t^2$

12. $4x^2y^2 - 121$ 14. $i_1^2 + 6i_1 + 9$ 16. $25a^2 + 20ab + 4b^2$ 18. $y^2 - 12y + 36$

20. $9x^2 + 60xy + 100y^2$ 22. $a^2 - 10ap + 25p^2$ 24. $9p^2 - 24pq + 16q^2$ 26. $y^2 - 3y - 40$

28. $t^2 - 8t + 7$ 30. $4x^2 - 12x - 7$ 32. $6y^2 - 5y + 1$ 34. $14s^2 + 47s + 30$

36. $24x^2 + 29xy - 4y^2$ 38. $5n^2 - 125$ 40. $16c^3 - 36c$ 42. $175r^3 + 140r^2b + 28rb^2$

44. $8p^3 - 112p^2 + 392p$ 46. $150t^4 - 180st^3 + 54s^2t^2$ 48. $x^2 + 9y^2 + 6xy + 4x + 12y + 4$

50. $2x^2 + 2y^2 - 4xy + 4x - 4y + 2$ 52. $8s^3 + 36s^2 + 54s + 27$ 54. $x^3 - 15x^2y + 75xy^2 - 125y^3$

56. $4a^2 - 4ac + c^2 - 4$ 58. $a^3 - 27$ 60. $8x^3 + 27a^3$ 62. $h^2L^2 + h^2L$ 64. $4J^2 + 4J - 3$

66. $4w - 4hw - h^2w + h^3w$ 68. $1 - z - z^2 + z^3$

Exercises 5-2, p. 156

2. $3(a - b)$ 4. $2(x^2 + 1)$ 6. $4s(s + 5)$ 8. $5a(a - 4x)$ 10. $3p^2(6p - 1)$

12. $5(2a - b + 3c)$ 14. $2q(2p - 7q - 8pq)$ 16. $3a(9ab - 8b - 3)$

18. $5a(1 + 2x - y + 4z)$ 20. $(r + 5)(r - 5)$ 22. $(7 + z)(7 - z)$

24. $(9z + 1)(9z - 1)$ 26. $(6s + 11t)(6s - 11t)$ 28. $(6ab + 13c)(6ab - 13c)$

30. $(a - b + 1)(a - b - 1)$ 32. $5(a + 5)(a - 5)$ 34. $4(x + 5y)(x - 5y)$

36. $a(x + 2 + y)(x + 2 - y)$ 38. $(y^2 + 9)(y + 3)(y - 3)$ 40. $2(x^2 + 2y^2)(x^2 - 2y^2)$

42. $\dfrac{5 - n}{1 + n}$ 44. $\dfrac{2k}{3k - 1}$ 46. $(a + c)(m + n)$ 48. $y(1 + 6y^2)(2 - y)$

50. $(x + 1)(x - 1)(x - 5)$ 52. $(2p + q)(2p - q + 1)$ 54. $d^2(4D^2 - 4dD - d^2)$

56. $K(1 + a)(1 - a)$ 58. $R(2I + 3i)(2I - 3i)$ 60. $\dfrac{R}{(T_2^2 + T_1^2)(T_2 + T_1)(T_2 - T_1)}$

Exercises 5-3, p. 162

2. $(x + 1)(x - 6)$ 4. $(a - 2)(a + 16)$ 6. $(r - 2)(r - 9)$ 8. $(y + 4)^2$ 10. $(b - 6c)^2$

12. $(2n + 1)(n - 7)$ 14. $(5x - 1)(x + 2)$ 16. $(y - 1)(7y - 5)$ 18. $(x - 1)(5x + 2)$

20. prime   22. $(3x + 7y)(x - 2y)$ 24. prime   26. $(4r - s)(r + 3s)$ 28. $(4q + 3)^2$

30. $(ac - 1)^2$ 32. $(3x - 4)(2x + 3)$ 34. $(6n - 5)(2n + 3)$ 36. $(2x - y)(6x + 5y)$

38. $(2r + s)(4r - 9s)$ 40. $3(y - 6)(2y + 1)$ 42. $2(6x - y)(x + 2y)$

44. $x^2(3x - 5)(2x - 1)$ 46. $(x - 3y + 2z)(x - 3y - 2z)$ 48. $(r + s - t)(r - s + t)$

50. $(x - 2)^3$ 52. $(x - 3)(x^2 + 3x + 9)$ 54. $3(p + 6)(p - 3)$ 56. $2x(x - 7)^2$

58. $b(T - 20)^2$ 60. $D(D - d)(D^2 + Dd + d^2)$

Exercises 5-4, p.166

2. $\dfrac{63}{45}$ 4. $\dfrac{4n^2x^3y}{6n^3x}$ 6. $\dfrac{7a + 14}{a^2 + a - 2}$ 8. $\dfrac{x^2 - 2x + 1}{x^2 - 1}$ 10. $\dfrac{5}{13}$ 12. $\dfrac{2a}{3a^3b^2}$

14. $\dfrac{x - 3}{3}$ 16. $\dfrac{2x + 5}{2x^2}$ 18. $\dfrac{2}{5}$ 20. $\dfrac{a}{3z^2}$ 22. $\dfrac{1}{t + a}$ 24. $\dfrac{r - 4s}{2r - s}$ 26. $\dfrac{x^2 - y^2}{x^2 + y^2}$

28. $5x$ 30. $\dfrac{3}{4t^2}$ 32. $\dfrac{2a + 3b}{2a}$ 34. $\dfrac{y(3y + 4)}{y + 4}$ 36. $\dfrac{2r + s}{3r - s}$ 38. $\dfrac{x - 2}{2}$

40. $x(2x + 1)$ 42. $\dfrac{6 + x}{3 - x}$ 44. $-(x + y)$ 46. $-\dfrac{3a + 2}{a + 1}$ 48. $\dfrac{(x - 7)(3x + 1)}{(3x + 2)(x + 7)}$

50. $x - 2$ 52. $a - 2$ 54. (a)   56. neither

58. $\dfrac{16(t - t_0)(t - t_0 - 3)}{3}$ 60. $\dfrac{r_0^2 + r_0 r_i + r_i^2}{r_0 + r_i}$

Exercises 5-5, p. 170

2. $\dfrac{13}{15}$ 4. $6ay^3$ 6. $\dfrac{13}{80}$ 8. $\dfrac{2r^2}{t^2}$ 10. $\dfrac{yz^2}{6(y - 2)}$ 12. $\dfrac{x + 2y}{x + y}$ 14. $\dfrac{a(a - 3)}{3(a - 1)}$

16. $-\dfrac{3x + 4}{x + 1}$ 18. $\dfrac{3(x - 3)}{x^2(x + 5)}$ 20. $\dfrac{x(x - 4)}{2(x - 2)}$ 22. $\dfrac{(x - 7)(2x + 3)}{x + 3}$ 24. $\dfrac{a(2a + 1)^2}{b(2b + 1)^2}$

26. $\dfrac{2(x - 2)(x + 2)}{(x - 3)(3x^2 + 8x - 4)}$ 28. $\dfrac{2x + 5}{x}$ 30. $\dfrac{u^3w}{90v^3}$ 32. $(2x + 1)(x + 4)(3 - x)$

34. $\dfrac{(x + 1)^3}{30}$ 36. $x(x + 1)(x^2 + 1)$ 38. $\dfrac{12wv^2}{g}$ 40. $\dfrac{8\ell u}{\pi a^4}$

10

2. $\dfrac{8}{13}$ 4. $\dfrac{5}{a}$ 6. $\dfrac{2}{9}$ 8. $\dfrac{t-6}{2a}$ 10. $\dfrac{2+3s}{s^2}$ 12. $\dfrac{ay^3-4b}{6y^4}$ 14. $\dfrac{2bc-24ac-9ab}{4abc}$

16. $\dfrac{19-3x}{4}$ 18. $\dfrac{20-3a}{12(2y+1)}$ 20. $\dfrac{3a-x-y}{a^2(x+y)}$ 22. $\dfrac{4x-1}{x(x+2)}$ 24. $\dfrac{-(3x+4)}{(x+2)^2}$

26. $\dfrac{2a+5}{3(a-1)(a+1)}$ 28. $\dfrac{10x^2+x-1}{2x^2(x-2)}$ 30. $\dfrac{6x^2-x+1}{(2x-1)(2x-5)(2x+3)}$

32. $\dfrac{-x^4+3x^3+x^2+6x}{x^2(2x-1)(x-1)(x+1)}$ 34. $x+1$ 36. $\dfrac{b(a+b)(a^2-5a-ab+8b)}{a(8a^2-b-8b^2)}$

38. $\dfrac{(x-1)(2x+1)}{4x(2-x)}$ 40. $\dfrac{3u+v}{4(u+v)}$ 42. $\dfrac{-6h}{(2x+2h-1)(2x-1)}$

44. $\dfrac{-2h(2x+h)}{(x^2+2hx+h^2+4)(x^2+4)}$ 46. $\dfrac{ry^2-x^3+rxy}{xy^2}$ 48. $\dfrac{a^4+a^3\ 2a^2+a+1}{a^2}$

50. $\dfrac{128T^3+9T^2-54P}{128T^3}$ 52. $\dfrac{sL+R}{s^2LC+sRC+1}$

2. -5 4. 10 6. $-\dfrac{37}{10}$ 8. $\dfrac{17}{19}$ 10. $\dfrac{1}{9}$ 12. $-\dfrac{1}{2}$ 14. $\dfrac{12}{7}$ 16. $\dfrac{6}{7}$ 18. $\dfrac{17}{2}$ 20. $\dfrac{37}{6}$

22. -6 24. $\dfrac{63}{8}$ 26. no solution   28. $\dfrac{2}{5}$ 30. $\dfrac{6h+1}{4}$ 32. $\dfrac{2}{5a+2}$ 34. $\dfrac{AIS-AMc}{I}$

36. $\dfrac{Kb}{a-K}$ 38. $\dfrac{8A+4w^2+\pi w^2}{4w}$ 40. $\dfrac{f}{x-f}$ 42. $\dfrac{24EID}{x^4-4Lx^3+6L^2x^2}$ 44. $\dfrac{1}{P}$ 46. 260 h

48. 18h   50. 900 km   52. 57.1 g

2. $-28x^3y+49xy^2$ 4. x^2-16z^2 6. $16x^2-24xy+9y^2$ 8. $y^2-12y+35$

10. $20a^2x^2+13ax-21$ 12. $24s^2-7st-6t^2$ 14. $7(x-4y)$ 16. $3a(x-2x^4-3)$

18. $(30+n)(30-n)$ 20. $(5s^2+6t)(5s^2-6t)$ 22. $(2x-3)^2$ 24. $(2x+9y)^2$

26. $(x-9)(x+5)$ 28. $(n-10)(n-1)$ 30. $(5x-3)(x+1)$ 32. $(9x+16)(x-1)$

34. $(4x+3y)(3x-4y)$ 36. $2a^2(x+2)(2x+9)$ 38. $(x-1)^3$ 40. $(x-5a)(x^2+5ax+25a^2)$

42. $a(y+1)(x-1)$ 44. $(y-4)(t+y+4)$ 46. $-\dfrac{3rt^7}{4s}$ 48. $\dfrac{x+1}{x+3}$

50. $\dfrac{2(x-3)}{x}$ 52. $\dfrac{3x+y}{4(x-2y)}$ 54. $\dfrac{x+2y}{(x+y)(2x-y)}$ 56. $-2(1+y)$ 58. $\dfrac{6a+5}{20a^3}$

60. $\dfrac{a^2-5a+8}{2a^2b}$ 62. $\dfrac{(y+1)(y-1)}{y(y+2)}$ 64. $\dfrac{2x^2-9x+3}{(2x-1)(2x+3)(2x-3)}$ 66. $\dfrac{7-9x}{5(x-4)}$

68. $\dfrac{3+10y-4y^2}{y^2(y+2)(y-4)}$ 70. 3   72. $\dfrac{2ac(a-bc+b^2)}{b(a+c)}$ 74. $\dfrac{3-a}{9a}$ 76. no solution

78. $\frac{1}{2}[(x+y)^2+(x-y)^2]=\frac{1}{2}[x^2+2xy+y^2+x^2-2xy+y^2]=x^2+y^2$

80. $4b^2+4bn\lambda-4b\lambda+n^2\lambda^2-2n\lambda^2+\lambda^2$ 82. $x(2x-5)^2$ 84. $(R+r)(R-r)$

86. $p(1-p)(a-b)^2$ 88. $A=(6+x)(2-x)$ 90. $\dfrac{mc}{c^2-p^2}$ 92. $\dfrac{kpV-RT}{k^2p^2}$

94. $\dfrac{40C^3r^3+9C^2Lr^2-3L^3}{24C^3r^3}$ 96. $\dfrac{2kAm-gm^2+2kAML}{2k^2}$ 98. $\dfrac{2RV(R+1)}{2R+1}$ 100. $\dfrac{55}{7}$

102. $\dfrac{a(1000-y)}{b+y}$ 104. $\dfrac{x^2(100I-20Ix+Ix^2-B)}{(10-x)^2}$ 106. 2.2 min   108. 7.0 h

110. 68.6 km/h   112. 12 km

2. $a=5,\ b=1,\ c=-9$ 4. $a=9,\ b=-12,\ c=2$ 6. not quadratic

8. $a=3,\ b=26,\ c=-40$ 10. $-20, 20$ 12. $-0.4, 0.4$ 14. $2, -3$ 16. $5, 6$

18. 0, 7   20. −4, 4   22. $\frac{4}{7}$, −1   24. $\frac{5}{2}$ (double root)   26. $\frac{3}{2}$, −$\frac{2}{3}$ 28. $\frac{1}{5}$, −$\frac{9}{2}$
30. $\frac{1}{2}$, 3   32. $\frac{1}{a}$, −$\frac{1}{a}$ 34. 0, $\frac{3}{4}$ 36. $\frac{4}{3}$ (double root)   38. 0, 4   40. 0, $\frac{1}{a+b}$
42. after 9 s   44. 64 km/h   46. −1, 1   48. 4, 5   50. −12, 40   52. 10 cm

Exercises 6-2, p. 193

2. −10, 10   4. −$\sqrt{15}$, $\sqrt{15}$ 6. −12, 8   8. 4 ± $\sqrt{10}$ 10. −2, 3   12. −6, 1
14. −5 ± $\sqrt{29}$ 16. 2, 6   18. −1, $\frac{3}{4}$ 20. $\frac{1}{3}$(−2 ± $\sqrt{13}$)   22. $\frac{1}{3}$(1 ± $\sqrt{3}$)

24. $\dfrac{-q \pm \sqrt{q^2 - 4pr}}{2p}$

Exercises 6-3, p. 197

2. −2, 3   4. −6, 1   6. −5 ± $\sqrt{29}$ 8. 2, 6   10. −1, $\frac{3}{4}$ 12. $\frac{1}{3}$(−2 ± $\sqrt{13}$)
14. $\frac{1}{3}$(1 ± $\sqrt{3}$)   16. −$\frac{7}{10}$, $\frac{9}{4}$ 18. $\frac{1}{2}$(2 ± $\sqrt{-10}$)   20. $\frac{1}{6}$(3 ± $\sqrt{15}$)   22. 0, 6

24. −$\frac{1}{5}$, $\frac{7}{5}$ 26. −1.20, 1.98   28. −1.76, 0.704   30. $\frac{1}{2}$(7 ± $\sqrt{25 - 4a}$ 32. $\dfrac{1 \pm \sqrt{4c^2 - 3}}{2(c^2 - 1)}$
34. (a) after 102 s, (b) after 1.83 s and 100 s   36. 1.90 mm   38. 352 m, 448 m
40. 11.1 h, 13.1 h

Exercises 6-4, p. 202

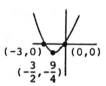
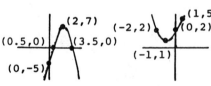

2.

4.

6.

8.

(−2,1)

(0,−3)

(0,−5)

(−2,−13)

$\left(-\frac{5}{4}, -\frac{25}{8}\right)$

(0,0)

(0,4) $\left(\frac{3}{2}, \frac{7}{4}\right)$

10.

12.

14.

16.

(−3,0)   (0,0)

$\left(-\frac{3}{2}, -\frac{9}{4}\right)$

(2,7)

(0.5,0)   (3.5,0)

(0,−5)

(1,5)

(−2,2)   (0,2)

(−1,1)

$\left(-\frac{1}{6}, \frac{1}{12}\right)$

(0,0)

(−1,−2)

(1,−4)

18. −2.2, 2.2   20. −0.5, 4.0   22. no real roots   24. −1.1, 7.7
26.   28.   30. 0.5 Ω, 7.8 Ω

16   L   d

47   32. 3.3 cm

15

400   q   8   t

Review Exercises for Chapter 6, p. 203

2. $-5, 2$ 4. $-3, 9$ 6. $\frac{1}{3}, \frac{3}{2}$ 8. $-1, -\frac{2}{3}$ 10. $-\frac{7}{6}, 5$ 12. $-8, \frac{1}{6}$ 14. $-6, 3$

16. $\frac{1}{2}(7 \pm \sqrt{53})$ 18. $-\frac{7}{3}, 2$ 20. $\frac{1}{10}(-7 \pm \sqrt{89})$ 22. $-0.295, 1.695$ 24. $1.57, -0.53$

26. $\frac{1}{2}(-3 \pm \sqrt{5})$ 28. $-4, \frac{7}{3}$ 30. $\frac{1}{12}(1 \pm \sqrt{-47})$ 32. $\frac{1}{8}(5 \pm \sqrt{153})$ 34. $\frac{1}{4}$ (double root)

36. $b \pm \sqrt{b^2 + 3b}$ 38. $1 \pm \sqrt{6}$ 40. $\frac{1}{2}(2 \pm \sqrt{7})$ 42. $2 \pm \sqrt{19}$ 44. -3, (5 is not a solution)

46.    48. 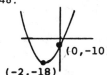   50. no real roots   52. $-0.4, 1.2$ 54. $94°C$

56. $15.5°$ 58. $195\ \Omega$

60. 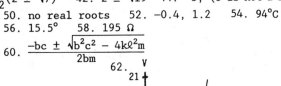 $\dfrac{-bc \pm \sqrt{b^2c^2 - 4k\ell^2 m}}{2bm}$ 62.

64. 0.5 m, 1.5 m   66. 450 km/h, 2000 km/h
68. 12.5 m   70. 0.5 m by 0.8 m   72. 4.6 cm

Exercises 7-1, p. 207

2. $+, -, -$ 4. $+, +, -$ 6. $+, +, -$ 8. $-, -, +$
10. $\sin \theta = \frac{1}{\sqrt{2}}$, $\cos \theta = -\frac{1}{\sqrt{2}}$, $\tan \theta = -1$, $\cot \theta = -1$, $\sec \theta = -\sqrt{2}$, $\csc \theta = \sqrt{2}$

12. $\sin \theta = -\frac{3}{5}$, $\cos \theta = \frac{4}{5}$, $\tan \theta = -\frac{3}{4}$, $\cot \theta = -\frac{4}{3}$, $\sec \theta = \frac{5}{4}$, $\csc \theta = -\frac{5}{3}$

14. $\sin \theta = -\frac{4}{5}$, $\cos \theta = -\frac{3}{5}$, $\tan \theta = \frac{4}{3}$, $\cot \theta = \frac{3}{4}$, $\sec \theta = -\frac{5}{3}$, $\csc \theta = -\frac{5}{4}$

16. $\sin \theta = \frac{40}{41}$, $\cos \theta = \frac{9}{41}$, $\tan \theta = \frac{40}{9}$, $\cot \theta = \frac{9}{40}$, $\sec \theta = \frac{41}{9}$, $\csc \theta = \frac{41}{40}$
18. III   20. IV   22. I   24. IV

Exercises 7-2, p. 214

2. $-\tan 89°$; $\sec 15°$ 4. $-\cos 10°$; $-\cot 70°$ $\sin 82°$; $\sec 45°$ 8. $\tan 20°$; $\csc 10°$
10. $-\tan 49° = -1.2$ 12. $\sin 76.6° = 0.973$ 14. $-\cos 18.82° = -0.9465$
16. $-\cot 43.47° = -1.055$ 18. 0.948 20. 1.936 22. 0.29 24. -6.449
26. $118.65°, 298.65°$ 28. $39.60°, 140.40°$ 30. $158.1°, 201.9°$ 32. $246.6°, 293.4°$
34. $223.0°$ 36. $348.9°$ 38. $304.93°$ 40. $108.04°$ 42. -0.907 44. -1.965 46. $>$
48. $<$ 50. 47.9 N   52. $120.4°$ 54. (a) -0.87, (b) -0.91 56. (a) 1.6, (b) -1.7

Exercises 7-3, p. 219

2. $\frac{\pi}{15}, \frac{5\pi}{4}$ 4. $\frac{\pi}{5}, \frac{7\pi}{4}$ 6. $\frac{4\pi}{3}, \frac{5\pi}{3}$ 8. $\frac{11\pi}{30}, \frac{35\pi}{18}$ 10. $54°, 150°$ 12. $84°, 240°$
14. $55°, 225°$ 16. $42°, 48°$ 18. 0.948 20. 1.82 22. 2.944 24. 1.50 26. $13.8°$
28. $97.57°$ 30. $1970°$ 32. $5730°$ 34. 0.8660 36. 0.9397 38. 1.732 40. 7.554
42. 0.5972 44. 0.341 46. 0.718 48. -10.7 50. $2.723, 3.561$ 52. $3.185, 6.240$
54. $1.077, 4.219$ 56. $0.2566, 2.885$ 58. 7.6×10^{-4} C   60. 0.0043 W/m^2

Exercises 7-4, p. 224

2. 47.2 cm   4. 0.690 m^2 6. 9.169 cm   8. 7920 m^2 10. $12\ 600$ rad/min
12. 2370 cm^2 14. 1.41 rad/s   16. 17.3 rad   18. 2930 cm/s   20. 238 m   22. 1.57 mm/s
24. 29 r/min   26. 20.42 cm   28. 7260 m/min   30. 1670 km/h   32. 1260 km/h   34. 24.1 m^2
36. 2.58 m^2 38. 1.745×10^{-5} 40. 0.119 m

Review Exercises for Chapter 7, p. 227

2. $\sin \theta = \frac{5}{13}$, $\cos \theta = -\frac{12}{13}$, $\tan \theta = -\frac{5}{12}$, $\cot \theta = -\frac{12}{5}$, $\sec \theta = -\frac{13}{12}$, $\csc \theta = \frac{13}{5}$

4. $\sin \theta = -\frac{3}{\sqrt{13}}$, $\cos \theta = -\frac{2}{\sqrt{13}}$, $\tan \theta = \frac{3}{2}$, $\cot \theta = \frac{2}{3}$, $\sec \theta = -\frac{\sqrt{13}}{2}$, $\csc \theta = -\frac{\sqrt{13}}{3}$

6. $-\sin 63°$, $-\cot 42°$ 8. $-\cos 77°$, $-\csc 80°$ 10. $\frac{\pi}{8}$, $\frac{9\pi}{5}$ 12. $\frac{3\pi}{20}$, $\frac{9\pi}{10}$ 14. 67.5°, 63°
16. 306°, 225°   18. 77.58°   20. 831°   22. 5.32   24. 2.5897   26. 0.3267   28. 6.726
30. 0.625   32. 0.070   34. −9.304   36. 0.8360   38. 0.971   40. −0.9528   42. −2.613
44. −0.2679   46. 1.037   48. −0.5176   50. 248.80°, 291.20°   52. 39.90°, 219.90°
54. 0.1047, 3.037   56. 1.232, 4.374   58. 120.72°   60. 164.78°   62. 15.7 cm
64. 131 cm   66. 108 cm²   68. 0.205 m²   70. 18.3 r/s   72. 19.6°   74. 172°   76. 12.2

Exercises 8-1, p. 234

2.(a) Scalar: only magnitude is specified, (b) vector: magnitude and direction are speci
4.(a) Scalar: only magnitude is specified, (b) vector: magnitude and direction are speci
6.   8.   10.   12.   14.   16.

18.   20.   22.   24.   26.   28.

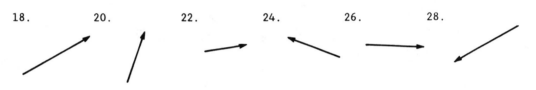

30.   32.   34.   36.

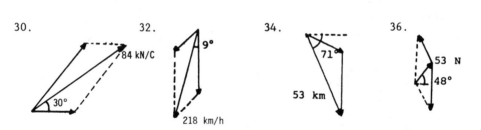

Exercises 8-2, p. 237

2. −194, 724   4. 458, −594   6. −4430, −8690   8. 0.0437, −0.0897
10. −15 000, 6540   12. 62.6, 25.9   14. −379.3, −340.0   16. 0.9525, −1.534
18. 3.2 m/s, −7.3 m/s   20. −0.243 N, 0.0588 N   22. 0.237 m, 1.18 m   24. 79.6 N

Exercises 8-3, p. 242

2. 623, 18.2° from A   4. 3725, 62.26° from A   6. 104, 329.9°   8. 758, 196.0°
10. 51.7, 232.9°   12. 9.926, 320.2°   14. 259, 68.8°   16. 24.4, 248.9°
18. 1.046, 2.31°   20. 159.23, 247.341°   22. 4160, 166.7°   24. 36.7, 348.5°

Exercises 8-4, p. 245

2. 29.5 N, $51.2°$ from 18.5-lb force   4. 28.0 kN/C, $15.9°$ below horizontal to right
6. 614 km, $54.8°$ south of west   8. 3220 km/h, $61.8°$
10. 38 900 kg·m/s, $13.8°$ from direction of car   12. 66.5 N
14. 3.0 km/h, $82.9°$ north of east   16. 9.8 m/s²   18. 5.20 km/h, perp. to bank
20. 41.3 N   22. 9.3 km/h, $11.6°$ downstream from barge heading
24. 3.97 A/m, $6.5°$ from vertical toward magnet

Exercises 8-5, p. 252

2. $B = 34°$, $a = 2.41$, $c = 4.75$ 4. $A = 79.5°$, $b = 29.1$, $c = 94.0$
6. $A = 61.11°$, $C = 49.52°$, $a = 3.391$ 8. $A = 35.64°$, $B = 67.93°$, $b = 239.2$
10. $A = 14.8°$, $a = 267$, $b = 896$ 12. $C = 57.3°$, $b = 14.6$, $c = 13.3$
14. $A = 89.83°$, $a = 1188$, $c = 1155$
16. $A_1 = 40.14°$, $B_1 = 124.24°$, $b_1 = 114.7$; $A_2 = 139.86°$, $B_2 = 24.52°$, $b_2 = 57.59$
18. $B_1 = 74.5°$, $C_1 = 48.4°$, $c_1 = 0.750$; $B_2 = 105.5°$, $C_2 = 17.4°$, $c_2 = 0.299$
20. $A = 90.0°$, $B = 60.0°$, $b = 17$ 22. $T_1 = 159$ N, $T_2 = 137$ N   24. 233.2 km
26. 7140 m   28. 28.3 m   30. 30.7 cm, $42.9°$
32. no solution; at least one measurement is in error

Exercises 8-6, p. 257

2. $B = 36.6°$, $C = 13.4°$, $a = 112$ 4. $A = 37.8°$, $C = 57.2°$, $b = 0.137$
6. $A = 48.65°$, $B = 61.39°$, $C = 69.96°$ 8. $A = 65.28°$, $B = 79.67°$, $C = 35.05°$
10. $B = 41.6°$, $C = 79.7°$, $a = 23.5$ 12. $A = 76.4°$, $B = 26.0°$, $c = 11.4$
14. $A = 55.45°$, $C = 10.19°$, $c = 10.59$ 16. $A = 82.36°$, $B = 44.49°$, $C = 53.15°$
18. $A = 13.6°$, $B = 127.0°$, $C = 39.4°$ 20. $A = 21°$, $B = 31°$, $C = 128°$ 22. 3.34 m
24. $74.21°$, $105.79°$ 26. 0.99 cm   28. $55.2°$ 30. $82.6°$ 32. 17.8 km

Review Exercises for Chapter 8, p. 258

2. $A_x = -6.90$, $A_y = 4.15$ 4. $A_x = 630.8$, $A_y = -184.0$ 6. 74, $23°$ with A
8. 93.69, $73.56°$ from A   10. 0.0118, $332.2°$ 12. 16 850, $96.08°$
14. 75.44, $146.19°$ 16. 6760, $160.2°$ 18. $B = 16.0°$, $a = 20.2$, $c = 14.4$
20. $C = 60.5°$, $a = 9.15$, $b = 7.25$ 22. $A = 28.28°$, $B = 91.07°$, $C = 60.65°$
24. $B = 78.60°$, $C = 24.34°$, $b = 12.14$ 26. no solution
28. $B = 42°$, $C = 82°$, $a = 920$ 30. $A = 14.9°$, $C = 59.7°$, $b = 0.779$
32. $B = 20.87°$, $C = 116.01°$, $c = 10.38$ 34. $A = 34.57°$, $B = 141.46°$, $C = 3.97°$
36. $A = 67.0°$, $B = 51.0°$, $C = 62.1°$ 38. 103.7 km/h, -135.4 km/h   40. 14 km/h
42. 9.8 km   44. $0.022°$ 46. 11.1 cm   48. 0.48 m   50. 368.61 m
52. 365 km, $22.0°$ south of west   54. 173 km from Halifax, 641 km from Boston
56. 1020 m/s, $18.7°$ from direction of plane, $5.5°$ below horizontal

Exercises 9-1, p. 264

2. -1, -0.7, 0, 0.7, 1, 0.7,
   0, -0.7, -1, -0.7, 0, 0.7,
   1, 0.7, 0, -0.7, -1

4. 0, 2.8, 4, 2.8, 0, -2.8,
   -4, -2.8, 0, 2.8, 4, 2.8,
   0, -2.8, -4, -2.8, 0

6.   **8.** 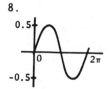 **10.**   **12.**

14. 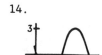 **16.** 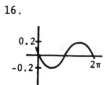 **18.**   **20.**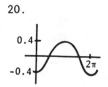

22. 0, −2.5, −2.7, −0.4, 2.3, **24.** 2, 1.1, −0.83, −2.0, −1.3,
2.9, 0.8, −2.0 0.57, 1.9, 1.5

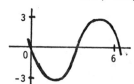 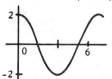

Exercises 9-2, p. 268

2. π **4.** $\frac{\pi}{5}$ **6.** $\frac{2\pi}{5}$ **8.** π **10.** $\frac{2}{3}$ **12.** $\frac{1}{5}$ **14.** 5π **16.** 8π **18.** 20   **20.** π^2

22.   **24.**   **26.**   **28.**   **30.**

32.   **34.**   **36.**   **38.**   **40.**

    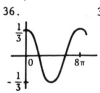   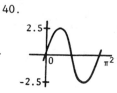

42. $y = \sin 5x$ **44.** $y = \sin \frac{\pi}{3}x$

46.   **48.**

 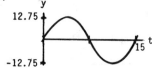

Exercises 9-3, p. 271

2. 3, 2π, $-\frac{1}{4}\pi$ 4. 2, 2π, $\frac{1}{8}\pi$ 6. 1, $\frac{2}{3}\pi$, $\frac{1}{6}\pi$ 8. 4, $\frac{2}{3}\pi$, $-\frac{1}{9}\pi$ 10. 2, 8π, -2π

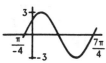

12. $\frac{1}{3}$, 4π, $\frac{1}{4}\pi$ 14. 2, 1, $\frac{1}{2}$ 16. 25, $\frac{2}{3}$, $-\frac{1}{6}$ 18. 1.8, 2, $-\frac{1}{3\pi}$ 20. 3, $\frac{1}{3}$, $\frac{1}{6\pi}$

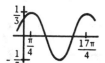   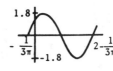   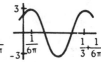

22. $\frac{1}{2}$, π, $\frac{1}{2\pi}$ 24. π, $2\pi^2$, $-\frac{1}{3}\pi$ 26.   28.

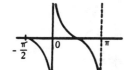   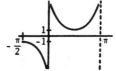      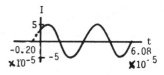

Exercises 9-4, p. 275

2. 0, −0.58, −1, −1.7, undef., 1.7,   4. −1, −1.2, −1.4, −2, undef., 2, 1.4,
   1, 0.58, 0, −0.58, −1, −1.7, undef.   1.2, 1, 1.2, 1.4, 2, undef.

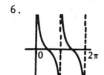   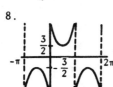

6.   8.   10.   12.

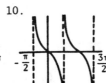   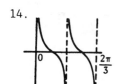

14.   16.   18.   20.

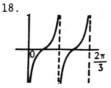   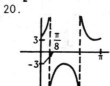   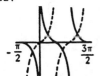   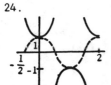

22.   24.   26.   28.

            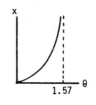

17

Exercises 9-5, p. 279

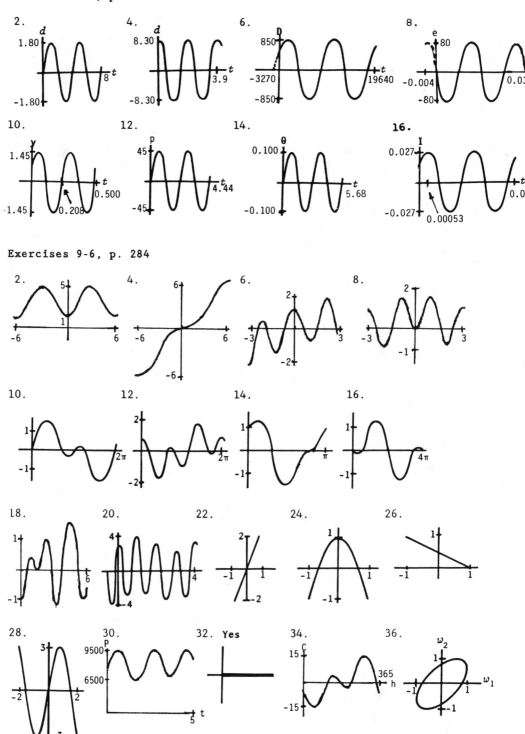

2.

4.

6.

8.

10.

12.

14.

16.

Exercises 9-6, p. 284

2.

4.

6.

8.

10.

12.

14.

16.

18.

20.

22.

24.

26.

28.

30.

32. **Yes**

34.

36.

Review Exercises for Chapter 9, p. 286

2.

4.

6.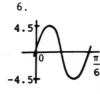

8.

10.

12.

14.

16.

18.

20.

22.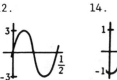

24.

26.

28.

30.

32.

34.

36.

38. $y = 2 \cos 2x$

40. $y = \sin(\frac{\pi}{4}x - \frac{\pi}{4})$

42.

44.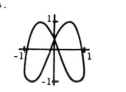

46.

48.

50.

52.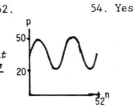

54. Yes

56.

58.

60.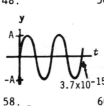

Exercises 10-1, p. 292

2. y^7 4. $\dfrac{1}{s^4}$ 6. $2(7^2)$ 8. $\dfrac{3^6}{4^9}$ 10. $\dfrac{27x^3}{y^6}$ 12. $\dfrac{t^2}{36s^4}$ 14. -1 16. 1 18. $\dfrac{1}{9x^2}$ 20. $\dfrac{7a}{x^3}$

22. $\dfrac{x^6}{9}$ 24. $\dfrac{n^4}{4m^2}$ 26. $\dfrac{b+a}{ab}$ 28. $\dfrac{1}{(3x+2y)^2}$ 30. $\dfrac{3^3}{7^4}$ 32. $\dfrac{3^3 s}{t}$ 34. $\dfrac{p^2}{n^2}$ 36. $\dfrac{1}{b^4}$

38. $\dfrac{s^6 - 1}{s^8}$ 40. $\dfrac{3a^3c^2 + z^5}{c^2z^6}$ 42. $-\dfrac{17}{12}$ 44. $\dfrac{b^2}{2ab^2 - 1}$ 46. 64   48. $\dfrac{8}{99}$ 50. $\dfrac{x + y}{xy}$

52. $2xy(y^2 - x^2)$ 54. $\dfrac{3x^2 - y - 2}{x^3}$ 56. $\dfrac{(2x - 1)(2x + 9)}{(x + 2)^2}$ 58. (a) 9^{-2}, (b) 3^{-4}

60. $303.551\,82 = 303.551\,82$ 62. $\dfrac{p[(1 + i)^n - 1]}{i(1 + i)^n}$ 64. $\dfrac{f_1 f_2}{f_1 + f_2 + d}$

Exercises 10-2, p. 297

2. 7   4. 3   6. 25   8. 32   10. $\dfrac{1}{2}$ 12. $\dfrac{1}{16}$ 14. 1   16. 81   18. 200   20. $\dfrac{3}{5}$

22. -2 24. -64 26. $\dfrac{21}{5}$ 28. $\dfrac{13}{10}$ 30. 82.608   32. 1.323   34. $x^{1/2}$ 36. $2r^{9/5}$

38. $\dfrac{1}{x^{3/2}}$ 40. $a^{19/10}$ 42. $\dfrac{4c^{4/3}}{b^{8/3}}$ 44. $\dfrac{1}{4x^2y^{8/5}}$ 46. $\dfrac{2x^2}{(x^3 + 1)^{1/3}}$ 48. $a^{1/12}$

50. $\dfrac{b^{1/6}}{a^{32/3}c^{4/3}}$ 52. $\dfrac{4a^{5/6}}{9b^{5/4}}$ 54. $\dfrac{a}{(a^2 - 1)^{1/4}}$ 56. $\dfrac{1 - 4x^2}{2x^3}$ 58. $\dfrac{12x + 1}{2x^{1/2}}$

60. $\dfrac{2 - 4x}{(3x - 1)^{2/3}}$ 62.   64.   66. 89.4   68. 229 km

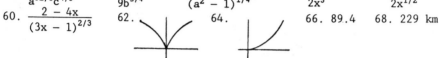

Exercises 10-3, p. 301.

2. $5\sqrt{6}$ 4. $7\sqrt{2}$ 6. $st^3\sqrt{s}$ 8. $xy^2z\sqrt{z}$ 10. $2b\sqrt{3a}$ 12. $3m^2n\sqrt{6mn}$ 14. $2\sqrt[4]{3}$ 16. $-2\sqrt[3]{}$

18. $a\sqrt[3]{5ab^2}$ 20. $xz^2\sqrt[5]{16y^3z}$ 22. $2\sqrt[7]{2}$ 24. $mn\sqrt{3n}$ 26. $\dfrac{1}{5}\sqrt{30}$ 28. $\dfrac{\sqrt{ab}}{b^2}$ 30. $\dfrac{1}{5}\sqrt[4]{250}$

32. $\dfrac{1}{2}\sqrt[6]{80}$ 34. $\sqrt{3}$ 36. $\sqrt[3]{3}$ 38. $200\sqrt{10}$ 40. $400\sqrt{10}$ 42. $\sqrt[3]{bc^2}$ 44. 2   46. $\sqrt[10]{3}$

48. $b^2\sqrt[4]{a}$ 50. $\dfrac{3}{4}\sqrt{2}$ 52. $\dfrac{\sqrt{xy(x^2 + y^2)}}{xy}$ 54. $\dfrac{\sqrt{x^2 - 4}}{x + 2}$ 56. $\sqrt{a^2 + b^2}$ 58. $3x - 1$

60. $\dfrac{1}{2}(1 + 2r)\sqrt{2}$ 62. $\sqrt{2ag}$ 64. $\dfrac{8Af\sqrt{f^2 + f_0^2}}{\pi^2(f^2 + f_0^2)}$

Exercises 10-4, p. 304

2. $5\sqrt{11}$ 4. $3\sqrt{6} - 2\sqrt{3}$ 6. $4\sqrt{7}$ 8. $-\sqrt{2}$ 10. $3\sqrt{3x} + 6\sqrt{2x}$ 12. $22\sqrt{3}$ 14. $5\sqrt{11}$

16. $-6\sqrt{3} - 6\sqrt{7}$ 18. $\dfrac{13}{7}\sqrt{21}$ 20. $\dfrac{2}{3}\sqrt{6} - 3\sqrt{2}$ 22. $\sqrt[3]{2}$ 24. $-\sqrt[12]{2}$ 26. $(x + 2y^2)\sqrt{2y}$

28. $-b\sqrt{15n}$ 30. $a(2 + 9b)\sqrt[5]{ab^4}$ 32. $\dfrac{(4x + 9y)\sqrt{6xy}}{12xy}$ 34. $\dfrac{\sqrt[4]{b^3c}}{b} - \sqrt[4]{bc}$

36. $\dfrac{(4 + x)\sqrt{x} - \sqrt{x(x - 1)}}{x}$ $(x \geq 1)$ 38. $-12\sqrt{10} = -37.947\,332$

40. $\dfrac{64}{7}\sqrt{14} = 34.209\,439$ 42. $-\dfrac{b}{a}$ 44. $\dfrac{hm(am + b)\sqrt{hm\pi}}{\pi}$

Exercises 10-5, p. 307

2. $\sqrt{102}$ 4. $7\sqrt{2}$ 6. 3   8. $5\sqrt[3]{10}$ 10. 27   12. $135\sqrt{5}$ 14. $2\sqrt{5}$ 16. $\dfrac{2}{7}\sqrt{7}$

18. $\sqrt{35} + \sqrt{10}$ 20. $15\sqrt{3} - 30$ 22. 33   24. $9 - 4\sqrt{5}$ 26. $21a + \sqrt{14a} - 4$

28. $-295 - 55\sqrt{6}$ 30. $3x - x\sqrt{3y}$ 32. $4x\sqrt{y} - 3\sqrt{2xy}$ 34. $4mn + 12n\sqrt{m} + 9n$

36. $24 + 14\sqrt{2} - 5\sqrt{35} - \sqrt{70}$ 38. $2\sqrt[4]{16}$ 40. $2\sqrt[10]{8x^7}$ 42. 2

44. $2a - 3\sqrt[6]{a^3b^2} + \sqrt[3]{b^2}$ 46. $1 + x - y - xy + \dfrac{(y - 1)\sqrt{xy}}{y}$ 48. $2a - \sqrt{6} - 2a$

50. $23 - \sqrt{35} = 17.083\,920$ 52. $282 - 183\sqrt{2} = 23.198\,918$ 54. $\dfrac{11 - 6x}{2\sqrt{3x} - 4}$

56. $\dfrac{4(x^2 - x + 1)}{\sqrt{x^2 + 1}}$ 58. $\dfrac{c}{a}$ 60. $\dfrac{100(2\sqrt{x} + 1)}{x}$

Exercises 10-6, p. 310

2. $\sqrt{21}$ 4. $\frac{1}{2}\sqrt{6}$ 6. $\frac{1}{2}\sqrt[4]{24}$ 8. $\sqrt[6]{54}$ 10. $\frac{2\sqrt[20]{32b^8}}{b}$ 12. $\frac{\sqrt{10}-2}{2}$ 14. $2\sqrt{x}+1$

16. $\sqrt{x}-\sqrt{2}$ 18. $\frac{2}{17}(6-\sqrt{2})$ 20. $\frac{2}{7}(2\sqrt{6}+\sqrt{10})$ 22. $-\frac{1}{6}(2\sqrt{7}+7)$ 24. $6\sqrt{5}+12$

26. $\frac{1}{3}(6\sqrt{2}-3\sqrt{5}-2\sqrt{10}+5)$ 28. $11\sqrt{15}-42$ 30. $-\frac{1}{93}(45+7\sqrt{66})$

32. $\frac{1}{11}(3-5\sqrt{3})$ 34. $\frac{12a+6b\sqrt{a}}{4a-b^2}$ 36. $\frac{6-12\sqrt{x}}{1-4x}$ 38. $6\sqrt{x}+3\sqrt{2x}+2\sqrt{2}+2$

40. $\frac{a\sqrt{1+a}+\sqrt{1-a^2}}{a^2+a-1}$ 42. $\frac{1}{2}(39+3\sqrt{143})=37.437\ 391$ 44. $\frac{3-16\sqrt{14}}{143}=-0.397\ 668\ 0$

46. $\frac{2}{\sqrt{19}+3}$ 48. $\frac{1}{2(\sqrt{3x+4}-\sqrt{3x})}$ 50. $\frac{2500-50\sqrt{V}}{2500-V}$ 52. $2Q\sqrt{\sqrt{2}-1}(\sqrt{2}+1)$

Review Exercises for Chapter 10, p. 312

2. $\frac{1}{2cz^2}$ 4. $-\frac{5y}{3}$ 6. 4   8. $\frac{1}{100}$ 10. $\frac{27}{8x^9}$ 12. 8   14. $\frac{a^8y^2}{x^4}$ 16. $\frac{9y^6}{x^4}$ 18. $\frac{a^7+1}{a^4}$

20. $\frac{6a^2}{1-2a^2}$ 22. $\frac{s^4}{(2+s^2t)^2}$ 24. $\frac{1}{x+y}$ 26. $\frac{2^8}{3^6b^3x^4}$ 28. $\frac{3-4x^2}{(1-x^2)^{1/2}}$ 30. $4\sqrt{6}$

32. $xy^2z^3\sqrt{x}$ 34. $2x^2y\sqrt{2x}$ 36. $2xy^2\sqrt{13y}$ 38. $\frac{3a\sqrt{5x}}{5x}$ 40. $\frac{1}{4}\sqrt{14}$ 42. $\frac{a^2\sqrt[3]{9a}}{b}$

44. $\frac{b\sqrt[10]{a^5b^2}}{a^2}$ 46. $\sqrt{17x}$ 48. $-10\sqrt{5}$ 50. $n(2m-n)\sqrt{n}$ 52. $(y-2)\sqrt[4]{2xy}$ 54. $40-8\sqrt{3}$

56. $15\sqrt{3}+30\sqrt{7}$ 58. $70+13\sqrt{6}$ 60. $4\sqrt{26}-9$ 62. $\frac{10a+5c\sqrt{a}}{4a-c^2}$ 64. $-\frac{4}{19}(3+2\sqrt{7})$

66. $\frac{2}{15}(7\sqrt{6}-18)$ 68. $\frac{6y-17\sqrt{yz}+5z}{4y-25z}$ 70. $\frac{\sqrt{a^2+b^2}}{ab}$ 72. $\frac{(a+b)\sqrt{ab}+a^2b^2(a+1)}{ab}$

74. $60-10\sqrt{30}=5.227\ 744\ 3$ 76. $\frac{214-25\sqrt{42}}{-29}=-1.792\ 464\ 9$ 78. 0.611 year   80. $km\sqrt{m(E-E_1)}$

82. $\frac{r^3v}{4r^3+3ar^2-a^3}$ 84. $\frac{ab}{(a^3+b^3)^{1/3}}$ 86. $\frac{\sqrt{LC_1C_2(C_1+C_2)}}{2\pi LC_1C_2}$ 88. $\frac{20(d-\sqrt{3d+400})}{d^2-3d-400}$

Exercises 11-1, p. 319

2. $11j$ 4. $-7j$ 6. $-0.1j$ 8. $4j\sqrt{3}$ 10. $-\frac{1}{3}j\sqrt{5}$ 12. $\frac{1}{3}j\sqrt{15}$ 14. $15;\ -15$
16. $-12;\ 12$ 18. j 20. 1   22. $3j$ 24. 4   26. $-6+8j$ 28. $-8-20j$ 30. $5+9j$
32. $-6-j$ 34. $2\sqrt{3}+3j\sqrt{3}$ 36. 9   38. $-3-2j$ 40. 6   42. $x=-3,\ y=4$
44. $x=-1,\ y=-8$ 46. $x=2,\ y=-2$ 48. $x=\frac{1}{2},\ y=-4$ 50. yes
52. pure imaginary

Exercises 11-2, p. 322

2. $-11-5j$ 4. $-0.23+0.86j$ 6. $7-22j$ 8. $-1+2j$ 10. $9+2j$ 12. $11-35j$
14. $3.3+8.8j$ 16. $-6-3j$ 18. $53-9j$ 20. $-195+260j$ 22. $-6\sqrt{6}$ 24. 1296
26. $8j\sqrt{6}$ 28. $10\sqrt{6}-9j\sqrt{5}$ 30. $6j+15j\sqrt{7}$ 32. $8-3j\sqrt{7}$ 34. $9+40j$ 36. $2-2j$
38. $\frac{2}{29}(3-7j)$ 40. $\frac{2}{13}(3+2j)$ 42. $\frac{1}{4}(8+9j)$ 44. $\frac{1}{28}(1+17j\sqrt{3})$ 46. $-\frac{1}{85}(16+13j)$
48. $\frac{1}{5}(1+3j)$ 50. $(-1-j)^2+2(-1-j)+2=1+2j-1-2-2j+2=0$
52. 10   54. $\frac{1}{5}(4-3j)$ 56. $\frac{1}{26}(89+23j)$ ohms   58. $(a+bj)(a-bj)=a^2+b^2$
60. $\frac{1}{j}=\frac{j}{j^2}=-j$

Exercises 11-3, p. 324

2.   4.   6. −2 + 5j   8. 2 − 3j   10. 2 − 2j   12. 5 − 3j

14. −4 + j   16. −1 + 2j   18. 7.5 + 3.5j   20. −4 − 10j   22. −3j   24. 2 + 4

26.   28.   30.   32.

neg.   conj.

conj.   neg.

•3+9j

−1−3j •

−3−9j •

6+3j
•2+j

−2−j •

Exercises 11-4, p. 328

2. 17(cos 241.9° + j sin 241.9°)   4. 13(cos 112.6° + j sin 112.6°)

6. 8.60(cos 324.5° + j sin 324.5°)   8. 6.51(cos 315.0° + j sin 315.0°)

10. 2(cos 315° + j sin 315°)   12. 11.38(cos 56.81° + j sin 56.81°)

14. 6(cos 0° + j sin 0°)   16. 2(cos 270° + j sin 270°)   18. −1.85 − 2.36j

22

20. $1.77 - 1.77j$ 22. $-12j$ 24. 15   26. $-200.3 + 92.86j$ 28. $0.485 - 0.875j$

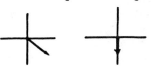 

30. $0.697 + 1.33j$ 32. $265.1 + 82.93j$ 34. -18.3 36. $-4625 - 197.1j$

38. 0.422 m, 20.5°   40. $3.62 + 0.971j$ μA

Exercises 11-5, p. 332

2. $5.00e^{2.36j}$ 4. $2.10e^{3.992j}$ 6. $16.72e^{-0.125j} = 16.72e^{6.16j}$ 8. $4650e^{5.70j}$

10. $0.0192e^{1.34j}$ 12. $827.6e^{1.921j}$ 14. $5.10e^{4.51j}$ 16. $6.08e^{0.165j}$ 18. $48.5e^{6.06j}$

20. $10\,170e^{2.573j}$ 22. $2.00(\cos 57.3° + j \sin 57.3°)$; $1.08 + 1.68j$

24. $2.50(\cos 220.0° + j \sin 220.0°)$; $-1.92 - 1.61j$

26. $0.800(\cos 171.9° + j \sin 171.9°)$; $-0.792 + 0.113j$

28. $820.7\underline{/200.1°} = -770.8 - 281.7j$ 30. $37.0e^{-1.14j}$ V/m   32. $3.17×10^{-4}e^{0.478j}$ 1/Ω

Exercises 11-6, p. 338

2. $15(\cos 165° + j \sin 165°)$ 4. $2.2\underline{/110°}$ 6. $3(\cos 150° + j \sin 150°)$ 8. $0.5\underline{/15°}$

10. $81(\cos 120° + j \sin 120°)$ 12. $1\underline{/340°}$ 14. $20.1\underline{/203.0°}$ 16. $579.4\underline{/72.37°}$

18. $7.62\underline{/336.8°} = 7 - 3j$ 20. $22.8\underline{/105.3°} = -6 + 22j$ 22. $1.10\underline{/74.1°} = \frac{1}{53}(16 + 56j)$

24. $3.81\underline{/246.8°} = -\frac{1}{2}(3 + 7j)$ 26. $16\underline{/0°} = 16$ 28. $125\underline{/339.4°} = 117 - 44j$

30. $3(\cos 40° + j \sin 40°)$, $3(\cos 160° + j \sin 160°)$, $3(\cos 280° + j \sin 280°)$

32. $2.00 + 3.00j$, $-2.00 - 3.00j$ 34. $2, -1 + j\sqrt{3}, -1 - j\sqrt{3}$

36. $0.924 + 0.383j, -0.383 + 0.924j, -0.924 - 0.383j, 0.383 - 0.924j$

38. $(1 + j)^2 = 1 + 2j + j^2 = 2j$ 40. $8.39\underline{/40.7°}$ cm

Exercises 11-7, p. 344

2. 10.1 V   4.(a) 2280 Ω, (b) 8.8°, (c) 13.1 V   6.(a) 54.5 Ω, $-34.4°$

8.(a) 47.3 Ω, (b) $-18.0°$ 10. 3.699 A   12. 86.7 Ω, 41.9°   14. 13.0 V   16. 0.106 mH

18. 421 Ω   20. 0.728 W

Review Exercises for Chapter 11, p. 346

2. $4 + 13j$ 4. $2 + 5j$ 6. $-28 + 44j$ 8. $1 + 8j$ 10. $\frac{4}{85}(9 + 2j)$ 12. $\frac{1}{17}(10 + 11j)$

14. $-\frac{1}{5}(6 + 7j)$ 16. $\frac{1}{25}(29 + 28j)$ 18. $x = \frac{9}{4}, y = \frac{3}{2}$ 20. $x = -\frac{9}{4}, y = \frac{3}{4}$

22. $2 + 2j$ 24. $4 + 7j$ 26. $5(\cos 36.9° + j \sin 36.9°) = 5e^{0.644j}$

28. $6.32(\cos 341.6° + j \sin 341.6°) = 6.32e^{5.96j}$

30. $363(\cos 154.2° + j \sin 154.2°) = 363e^{2.69j}$

32. $4(\cos 270° + j \sin 270°) = 4e^{4.71j}$ 34. $2.00 + 3.46j$

36. $1.069 - 2.168j$ 38. $-19 + 6.8j$

40. $-1.636 - 0.4189j$ 42. $-0.888 - 0.460j$

44. $43.22 - 10.45j$ 46. $20(\cos 277° + j \sin 277°)$ 48. $3.407\underline{/16.72°}$

50. $4.5(\cos 211° + j \sin 211°)$ 52. $25\underline{/246°}$ 54. $27.8\underline{/81.5°}$ 56. $12.22\underline{/352.34°}$

58. $3^6(\cos 216° + j \sin 216°)$ 60. $825\underline{/161.2°}$

62. $2^{10}\sqrt{2}(\cos 105° + j \sin 105°) = 2^9[(1 - \sqrt{3}) + (1 + \sqrt{3})j] = -375 + 1400j$

64. $\frac{1}{2^8}(\cos 240° + j \sin 240°) = -\frac{1}{2^9}(1 + j\sqrt{3})$ 66. $1, \frac{1}{2}(-1 + j\sqrt{3}), \frac{1}{2}(-1 - j\sqrt{3})$

68. $2(\cos 36° + j \sin 36°), 2(\cos 108° + j \sin 108°), -2, 2(\cos 252° + j \sin 252°),$
    $2(\cos 324° + j \sin 324°)$

70. $7.48 \ \Omega$ 72. $10.63 \ \mu F$ or $15.22 \ \mu F$ 74. $\sqrt{(A+B)^2\cos^2\omega t + (A-B)^2\sin^2\omega t}$, $\tan \theta = \frac{(A-B)\text{si}}{(A+B)\text{cc}}$

76. 2920 km/h, $31.0°$ 78. $\dfrac{k^2\mu^2 + \mu^2 - 2k\mu j - 1}{k^2\mu^2 + \mu^2 + 2\mu + 1}$ 80. $(e^{j\pi})^{1/2} = e^{j\pi/2} = \cos \frac{1}{2}\pi + j \sin \frac{1}{2}\pi =$

Exercises 12-1, p. 352

2. 6561 4. $\frac{1}{3}$ 6. $\log_5 25 = 2$ 8. $\log_8 64 = 2$ 10. $\log_3(\frac{1}{9}) = -2$ 12. $\log_{12} 1 = 0$

14. $\log_{81} 27 = \frac{3}{4}$ 16. $\log_{1/2} 4 = -2$ 18. $121 = 11^2$ 20. $1 = 15^0$ 22. $16 = 8^{4/3}$

24. $\frac{1}{8} = 32^{-3/5}$ 26. $\frac{1}{49} = 7^{-2}$ 28. $3 = (\frac{1}{3})^{-1}$ 30. 3 32. $-\frac{1}{2}$ 34. 512 36. $\frac{1}{49}$

38. 5 40. 8 42. 1.3 44. 16 46. -3 48. $\frac{1}{2}$ 50. $R = \log_{10}\left(\frac{I}{I_0}\right)$

52. $w = w_0 e^{-v/u}$ 54. $t = -\frac{1}{a}\log_e\left(1 - \frac{q}{q_0}\right)$ 56. $\frac{y}{2} = \log_5 x \to x = 5^{y/2} \to y = 5^{x/2}$

Exercises 12-2, p. 355

2. 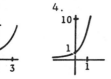 4. 6. 8. 10.

12. 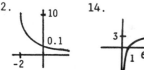 14. 16. 18. 20.

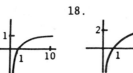

22. 24. 26. $V = 5000(1.0075)^{12t}$ 28. 30.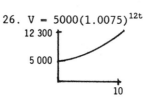

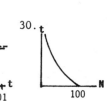

32. For $0 < b < 1$: $\log_b x > 0$ for $0 < x < 1$, $\log_b x < 0$ for $x > 1$

Exercises 12-3, p. 360

2. $\log_3 7 + \log_3 y$ 4. $\log_3 r - \log_3 s$ 6. $5 \log_8 n$ 8. $\log_2 x + \log_2 y - 2 \log_2 z$

10. $\frac{1}{7}\log_4 x$ 12. $\frac{1}{3}\log_3 y - \log_3 8$ 14. $\log_2 3x$ 16. $\log_8\left(\frac{6}{a}\right)$ 18. $\log_4 3^5$ 20. $\log_b\left(\frac{\sqrt{a}}{25}\right)$

22. -4 24. 0.1 26. $\frac{1}{3}$ 28. $\frac{2}{3}$ 30. $2 + \log_5 3$ 32. $-2 + \log_{10} 5$ 34. $\frac{1}{3}(3 + \log_2 3)$

24

36. $4 + 5 \log_7 3$ 38. $2 + 4 \log_{10} 2$ 40. $-3 + 2 \log_5 2$ 42. $y = 6x$ 44. $y = \dfrac{7}{x^2}$

46. $y = 100x^{3/2}$ 48. $y = \dfrac{1}{3}x^{4/3}$ 50. $y = \dfrac{4}{x^3}$ 52. $y = \dfrac{\sqrt{x}}{64}$ 54. 1.301 56. 3.903

58.

    60. $N = 10 \log_{10} \dfrac{R_1 I_1^2}{R_2 I_2^2}$

Exercises 12-4, p. 364

 2. 1.782 4. -3.247 6. 15.504 8. -7.0946 10. -2.0592 12. $5.217\ 09$
14. 0.0885 16. -0.1789 18. 8.49 20. 1.050×10^{-7} 22. 6.6736×10^8 24. 2.762×10^{-3}
26. $1.886\ 34 \times 10^{10}$ 28. 4.6156×10^{-11} 30. 98.5 32. 2.05 34. 3.18×10^{-158}
36. 9.07×10^{100} 38. 5.663 40. -5.3 42. 4.88 44. 0.201 A

Exercises 12-5, p. 368

 2. 6.447 4. 3.8228 6. -2.9386 8. $-7.720\ 86$ 10. 6.426 12. 5.971 14. 1.487
16. 1.785 18. 5.680 20. 4.1839 22. -6.1725 24. $-9.707\ 79$ 26. 9.0785
28. $-6.550\ 22$ 30. 30.6 32 1.88 34. $0.052\ 751$ 36. $0.991\ 99$ 38. $y = \dfrac{5e}{x^2}$

40. 34.7 years   42. 0.813 cm   44. $t = -0.1 \ln\left(1 - \dfrac{i}{0.6}\right)$

Exercises 12-6, p. 372

 2. -4 4. 1.511 6. -0.1042 8. -2.864 10. 1.431 12. 0.982 14. -9.015
16. 8.242 18. $\dfrac{1}{8}$ 20. 0.9739 22. 0.2 24. 1.58 26. 250 28. $4e$ 30. 64.5
32. no solution   34. 3 36. 2 38. 3.32 years   40. 0.459 s
42. $10 \log\left(\dfrac{I_1}{I_0}\right) - 10 \log\left(\dfrac{I_2}{I_0}\right) = d;\ 10 \log\left(\dfrac{I_1}{I_2}\right) = d;\ I_1 = I_2 10^{d/10}$

44. $10^{0.4} = 2.51$ 46. $q = q_0 e^{-t/RC}$ 48. $\dfrac{v^2}{2g}$ 50. $-5.0,\ 2.2$ 52. 3.3 s

Exercises 12-7, p. 376

2.   4.   6.   8.   10.

12.   14.   16.   18.   20.

    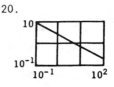

22.   24.   26.(a)   (b)   28.

  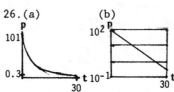

25

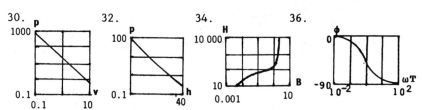

30.   32.   34.   36.

Review Exercises for Chapter 12, p. 379

2. 729   4. $\frac{1}{2}$ 6. 2   8. $\frac{3}{2}$ 10. 3   12. 16   14. $\log_5 7 - \log_5 a$ 16. $\frac{1}{2}\log_6 5$

18. $2 + \log_7 2$ 20. $-2 + \log_6 5$ 22. $1 + \frac{1}{2}\log_6 2 + \frac{1}{2}\log_6 y$ 24. $7 + 3\log_3 2$

26. $y = \sqrt{7x}$ 28. $\dfrac{2^{8/3}}{x^{1/2}}$ 30. $y - 1225x$ 32. $y - \dfrac{3}{x^2}$

34.   36.   38.   40.

42. 0.362   44. 6.00×10^{115} 46. 3.497   48. −0.6125   50. 1.25   52. 8.755   54. −0.2   56.

58.   60.   62. 15   64. 45   66. $e^{(P-D)/wa^2}$ 68. $V = 500(1.025)^4$

70.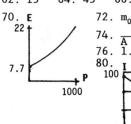

72. $m_0 = m_s e^{(v_m + gt_f)/u}$

74. $\dfrac{A}{A + 1}$

76. 1.903 dB   78. 1.40

80.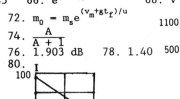

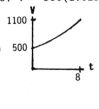

Exercises 13-1, p. 386

2. $x = 1.6$, $y = 0.8$; $x = -3.1$, $y = -13.3$ 4. $x = 2.7$, $y = 2.2$; $x = -0.7$, $y = -8.2$
6. no solution   8. $x = 1.4$, $y = -4.0$; $x = -1.4$, $y = -4.0$
10. $x = -1.4$, $y = 2.7$; $x = 1.4$, $y = 2.7$ 12. $x = -0.8$, $y = 4.8$
14. $x = 2.6$, $y = 1.6$; $x = 2.6$, $y = -1.6$; $x = -2.6$, $y = 1.6$; $x = -2.6$, $y = -1.6$
16. $x = 4.0$, $y = 3.0$; $x = 4.0$, $y = -3.0$; $x = -4.0$, $y = 3.0$; $x = -4.0$, $y = -3.0$
18. $x = 0.5$, $y = 1.8$; $x = 4.2$, $y = -0.9$ 20. $x = -1.9$, $y = 0.3$; $x = 0.9$, $y = 1.8$
22. $x = 0.4$, $y = -2.0$; $x = 2.3$, $y = 1.6$ 24. $x = 1.3$, $y = 0.3$ 26. 360 m, 210 m
28. 1.7 cm, 3.2 cm

Exercises 13-2, p. 391

2. $x - 1$, $y - 1$; $x - -1$, $y - -3$ 4. $x - 3$, $y - 4$; $x - -4$, $y - -3$
6. $x - 1$, $y - 1$; $x - -5$, $y - 7$ 8. $x - -6$, $y - 0$; $x - \frac{66}{13}$, $y - \frac{24}{13}$
10. $x - 1$, $y - 2$; $x - 2$, $y - 1$ 12. $x - 1$, $y - -4$, $x - -2$, $y - 2$
14. $x - 1$, $y - 0$; $x - -1$, $y - 0$; $x - \sqrt{3}$, $y - 2$; $x - -\sqrt{3}$, $y - 2$
16. $x - 0$, $y - 5$; $x - -3$, $y - -4$; $x - 3$, $y - -4$ 18. $x - -\frac{1}{2}$, $y - \frac{1}{2}\sqrt{10}$; $x - -\frac{1}{2}$, $y - -\frac{1}{2}\sqrt{10}$
20. $x - \sqrt{2}$, $y - \sqrt{2}$; $x - \sqrt{2}$, $y - -\sqrt{2}$; $x - -\sqrt{2}$, $y - \sqrt{2}$; $x - -\sqrt{2}$, $y - -\sqrt{2}$
22. $x - 2$, $y - 5$; $x - 2$, $y - -5$; $x - -2$, $y - 5$, $x - -2$, $y - -5$
24. $x - \sqrt{3}$, $y - 0$; $x - -\sqrt{3}$, $y - 0$ 26. $x - 50$ km, $h - 25$ km
28. $x = 1.25$ Ω, $R = 1.56$ Ω   30. 5.03 cm, 7.03 cm   32. $p = \$350$, $t = 100$ tables

Exercises 13-3, p. 395

2. $-4, -2, 2, 4$ 4. $-\frac{1}{2}, \frac{1}{2}, -2j, 2j$ 6. $-2, 5$ 8. $-1, 1$ 10. $\frac{1}{16}$ 12. 256

14. 4 16. $1, \frac{27}{8}$ 18. $-\frac{217}{216}, 0$ 20. $\frac{1}{6}(-9 \pm \sqrt{141}), \frac{1}{2}(-3 \pm \sqrt{5}$ 22. $0, \sqrt{2}, -\sqrt{2}$

24. $-3, 2, \frac{3}{2}(-1 + j\sqrt{3}), -\frac{3}{2}(1 + j\sqrt{3}), 1 + j\sqrt{3}, 1 - j\sqrt{3}$ 26. $\sqrt{\dfrac{2C}{\sqrt{B^2 - 4AC + 4\mu C} - B}}$

28. 5 cm, 25 cm, 25 cm

Exercises 13-4, p. 398

2. 5 4. 4 6. $\frac{3}{2}$ 8. 10 10. 4 12. -11 14. 1 16. 1, 9 18. 10

20. 9 22. $3 - 2\sqrt{6}$ 24. 2 26. 12 28. 4 30. 6 32. 4 34. $\dfrac{2nf - mv^2}{2}$

36. $\sqrt{r_2^2 + \left(kC + A - \sqrt{R_1^2 - R_2^2}\right)^2}$ 38. 1.33 m, 11.3 m 40. 5.1 m

Review Exercises for Chapter 13, p. 399

2. $x = 4.7, y = -1.7; x = -1.7, y = 4.7$ 4. no solution
6. $x = 1.3, y = 0.8; x = 1.3, y = -0.8; x = -1.3, y = 0.8; x = -1.3, y = -0.8$
8. $x = 1.2, y = -1.7$ 10. $x = 2.2, y = 0.8$ 12. $x = 1, y = 1$
14. $x = -1, y = 1; x = 1, y = 1$ 16. $x = -1, y = 1; x = \frac{11}{9}, y = -\frac{1}{9}$
18. no solution 20. $x = 2, y = 3; x = \frac{30}{17}, y = 5$
22. $-1, 3, \frac{3}{2}(-1 + j\sqrt{3}), -\frac{3}{2}(1 + j\sqrt{3}), \frac{1}{2}(1 + j\sqrt{3}), \frac{1}{2}(1 - j\sqrt{3})$ 24. 256
26. $2, -2, \frac{1}{3}j, -\frac{1}{3}j$ 28. $\frac{1}{4}$ 30. $-6, 1, \frac{1}{2}(-5 + \sqrt{21}), -\frac{1}{2}(5 + \sqrt{21})$ 32. 29 34. $\frac{4}{3}$
36. $\frac{1}{2}(4 + \sqrt{3}), \frac{1}{2}(4 - \sqrt{3})$ 38. no solution 40. 9 42. $\dfrac{1}{\omega(\omega L - R)}$
44. $\dfrac{2ec}{eZ + \sqrt{e^2Z^2 + 4cV}}$ 46. $\dfrac{m_1(u_1 - v_1)}{u_1 + v_1}$ 48. 1.0 m, 2.2 m 50. 1.12 cm
52. 2.52 cm, 4.48 cm or 0.863 cm, 6.14 cm 54. 10 cm 56. 0.035 h, 0.115 h

Exercises 14-1, p. 403

2. -4 4. 77 6. 2 8. 172 10. -35 12. 323 14. 606 16. 79 18. no
20. yes 22. yes 24. no 26. yes 28. yes
30. $(3x^3 - x^2 + 6x - 2) \div (x^2 + 2) = 3x - 1$; no 32. -2

Exercises 14-2, p. 408

2. $x^2 - x - 3, R = -4$ 4. $x^3 - 7x^2 + 20x - 59, R = 77$ 6. $x^2 + 9x + 20, R = 2$
8. $2x^3 - 8x^2 + 22x - 58, R = 172$ 10. $2x^2 - 8x + 17, R = -35$
12. $4x^3 + 12x^2 + 35x + 110, R = 323$ 14. $x^3 - 9x^2 + 37x - 150, R = 606$
16. $3x^3 + 3x^2 + 15x + 15, R = 79$ 18. $x^4 + 3x^3 - 3x^2 + 3x - 3, R = -5$
20. $20x^3 - 44x^2 + 32x - 28, R = 0$ 22. no 24. no 26. no 28. yes
30. no 32. yes 34. yes 36. no

Exercises 14-3, p. 413

(Note: Unknown roots listed.)
2. $j, -j$ 4. $\frac{1}{2}(-1 \pm j\sqrt{3})$ 6. $-\frac{1}{2}, 5$ 8. $-1, -2 - j$ 10. $\frac{1}{2}(-1 \pm \sqrt{13})$
12. $-\frac{1}{2}, -\frac{1}{2}$ 14. $\frac{1}{2}(-1 \pm \sqrt{5})$ 16. $-3j, 9, -1$ 18. $j, -j$ 20. $2j, -2j$
22. $-\frac{1}{2}j, -\frac{1}{2}(1 + j\sqrt{3}), \frac{1}{2}(-1 + j\sqrt{3})$ 24. $-1, 2, j, -j$

Exercises 14-4, p. 419

2. $-3, 1, 1$ 4. $-1, \frac{1}{2}(1 \pm j\sqrt{3})$ 6. $-\frac{1}{2}, \frac{1}{2}(1 \pm \sqrt{5})$ 8. $-2, \frac{1}{4}, 3$
10. $-2, 2, \frac{1}{2}(-1 \pm j\sqrt{7})$ 12. $-1, 2, 2j, -2j$ 14. $-\frac{1}{2}, -1, -1, -1$
16. $\frac{2}{3}, -\frac{1}{3}, 2j, -2j$ 18. $\sqrt{2}. \sqrt{3}. -\sqrt{2}, -\sqrt{3}, 2j, -2j$ 20. $-2, -1, -1, -\frac{1}{2}, 2$
22. 2.56 24. -1.56 26. 20 cm   28. a   30. 1.3 s, 4.7 s   32. 3 cm, 4 cm, 5 cm
34. 1.23 cm or 2.14 cm   36. $2.30, -1.30$

Review Exercises for Chapter 14, p. 420

2. -7 4. 61 6. yes   8. no   10. $3x^2 + 7x + 21$, $R = 70$ 12. $3x^2 - 17x + 75$, $R = -30$
14. $x^3 - 3x^2 - 9x - 26$, $R = -86$ 16. $x^5 - 4x^4 + 16x^3 - x^2 + 9x - 45$, $R = 172$
18. yes   20. no   (Note: unlisted roots given for 22-32.)   22. $\frac{1}{4}(1 \pm \sqrt{17})$
24. $j, -j$ 26. $-2, 3$ 28. $-1 - j, \sqrt{2}, -\sqrt{2}$ 30. $\frac{1}{2}j\sqrt{2}, -\frac{1}{2}j\sqrt{2}$ 32. $1 - j, -1 \pm j$
34. $2, 2, 4$ 36. $-2, \frac{1}{2}, 3$ 38. $-3, -\frac{1}{2}, \frac{1}{3}$ 40. $-3, -\frac{5}{2}, 1, 2$ 42. 2 44. $\frac{1}{2}, 3$
46. -0.59 48. \$3   50. 10.0 mm   52. $3\ \mu F, 4\ \mu F, 6\ \mu F$ 54. 8.0 m, 8.0 m, 2.5 m
56. 2.7 mm, 1.7 mm

Exercises 15-1, p. 428

2. -116 4. -66 6. 39 8. 17.777 10. 71 12. 132 14. -65 16. 4
18. $x - \frac{1}{2}, y - -2, z - 1$ 20. $x - 4, y - \frac{2}{3}, z - -2$
22. $x - \frac{1}{2}, y - 2, z - 1, t - -1$ 24. $x - 2, y - -1, z - 3, t - 1$ 26. $0.618, 1.618$
28. 40 g, 30, g, 20 g, 10 g

Exercises 15-2, p. 434

2. 72 4. -150 6. 0 8. 0 10. 88 12. 90 14. -26 16. 0 18. 450
20. -1032 22. $x - 2, y - -2, z - 3$ 24. $x - \frac{18}{7}, y - -\frac{16}{7}, z - -\frac{5}{7}$
26. $x - -1, y - -2, z - 4, t - \frac{1}{2}$ 28. $x - \frac{1}{3}, y - -2, z - 2, t - 1$
30. $A - 250$ N, $B - 600$ N, $C - 1000$ N, $D - 750$ N   32. $12, 10, 6, 4$

Exercises 15-3, p. 439

2. $x = -2, y = 7, z = -9, r = 4, s = 4, t = -5$ 4. $a = 0, b = 5, c = 3, d = 0, e = -6, f =$
6. $x = 5, y = -1$ 8. matrices not equal; different number of columns
10. $\begin{pmatrix} 5 & -1 & 16 \\ 5 & -5 & -5 \end{pmatrix}$ 12. $\begin{pmatrix} -0.2 & -7.5 & -11.7 \\ -3.4 & 5.5 & 7.4 \\ 3.7 & 10.8 & 4.3 \end{pmatrix}$ 14. $\begin{pmatrix} -2 & -1 & -1 & -3 \\ -2 & -5 & -9 & 4 \end{pmatrix}$ 16. cannot be added
18. $\begin{pmatrix} 1 & 14 & -19 & 6 \\ 10 & -8 & 15 & -2 \end{pmatrix}$ 20. $\begin{pmatrix} -4 & 7 & -15 & -3 \\ 2 & -17 & -11 & 8 \end{pmatrix}$ 22. $A + 0 = \begin{pmatrix} -1 & 2 & 3 & 7 \\ 0 & -3 & -1 & 4 \\ 9 & -1 & 0 & -2 \end{pmatrix} = A$
24. $3(A + B) = 3A + 3B = \begin{pmatrix} 9 & 3 & 0 & 21 \\ 15 & -9 & -6 & 15 \\ 30 & 30 & 24 & 0 \end{pmatrix}$ 26. $I_1 = -6$ A, $I_2 = 4$ A, $I_3 = 2$ A
28. $A - 2B = \begin{pmatrix} 5 & 0 & 20 \\ 6 & 10 & 24 \end{pmatrix}$

Exercises 15-4, p. 444

2. $(4 \quad -12)$ 4. $\begin{pmatrix} 71 & -38 \\ -45 & 50 \end{pmatrix}$ 6. $\begin{pmatrix} 11 & 0 \\ 35 & 20 \end{pmatrix}$ 8. $\begin{pmatrix} -2802 & 1619 & -3940 \\ -113 & 236 & -280 \\ -552 & 454 & -872 \end{pmatrix}$
10. $\begin{pmatrix} 29 & -22 \\ -4 & 14 \\ 0 & 5 \end{pmatrix}$ 12. $\begin{pmatrix} 31 \\ 3 \end{pmatrix}$ 14. $AB - \begin{pmatrix} 14 & -12 \\ 7 & 29 \end{pmatrix}$, $BA - \begin{pmatrix} 6 & -4 & 0 \\ -18 & 32 & -30 \\ -14 & 6 & 5 \end{pmatrix}$
16. AB not defined, $BA - (-11 \quad 15 \quad 23)$ 18. $AI - IA - A$ 20. $AI - IA - A$
22. $B \neq A^{-1}$ 24. $B - A^{-1}$ 26. yes   28. no   30. $N^2 - \begin{pmatrix} 0 & -1 \\ 1 & 0 \end{pmatrix}\begin{pmatrix} 0 & -1 \\ 1 & 0 \end{pmatrix} - \begin{pmatrix} -1 & 0 \\ 0 & -1 \end{pmatrix} - -I$

32. $J^2 = \begin{pmatrix} -1 & 0 \\ 0 & -1 \end{pmatrix} = -I$, $J^3 = \begin{pmatrix} -j & 0 \\ 0 & -j \end{pmatrix} = -J$, $J^4 = \begin{pmatrix} 1 & 0 \\ 0 & 1 \end{pmatrix} = I$ 34. $\begin{pmatrix} -2.464 \\ 3.732 \\ 0 \end{pmatrix}$

36. $X = 1.0x + 0.1y$, $Y = 0.5x + 1.0y + 0.1z$, $Z = 0.3x + 0.4y + 1.0z$

Exercises 15-5, p. 449

2. $\begin{pmatrix} -\frac{2}{3} & -1 \\ -1 & -2 \end{pmatrix}$ 4. $\begin{pmatrix} \frac{5}{44} & -\frac{1}{44} \\ -\frac{1}{11} & -\frac{2}{11} \end{pmatrix}$ 6. $\begin{pmatrix} 1 & 1 \\ 3 & \frac{7}{2} \end{pmatrix}$ 8. $\begin{pmatrix} 0.125 & -0.079 \\ -0.028 & -0.157 \end{pmatrix}$ 10. $\begin{pmatrix} -4 & -5 \\ 1 & 1 \end{pmatrix}$

12. $\begin{pmatrix} \frac{2}{5} & \frac{3}{5} \\ \frac{3}{10} & \frac{1}{5} \end{pmatrix}$ 14. $\begin{pmatrix} -5 & 3 \\ -3 & 2 \end{pmatrix}$ 16. $\begin{pmatrix} -\frac{5}{16} & \frac{3}{16} \\ -\frac{7}{16} & \frac{1}{16} \end{pmatrix}$ 18. $\begin{pmatrix} 10 & -2 & 3 \\ -5 & 1 & -2 \\ -1 & 0 & -1 \end{pmatrix}$ 20. $\begin{pmatrix} 67 & 22 & -9 \\ -15 & -5 & 2 \\ -6 & -2 & 1 \end{pmatrix}$

22. $\begin{pmatrix} -31 & -12 & 5 \\ 6 & 2 & -1 \\ 7 & 3 & -\frac{1}{2} \\ 2 & 2 & \end{pmatrix}$ 24. $\begin{pmatrix} \frac{1}{5} & \frac{5}{12} & \frac{7}{20} \\ \frac{1}{5} & \frac{1}{6} & \frac{1}{10} \\ \frac{1}{5} & -\frac{1}{6} & \frac{1}{10} \end{pmatrix}$ 26. $\begin{pmatrix} -31 & -12 & 5 \\ 6 & 2 & -1 \\ 7 & 3 & -\frac{1}{2} \\ 2 & 2 & \end{pmatrix}$ 28. $\begin{pmatrix} \frac{1}{5} & \frac{5}{12} & \frac{7}{20} \\ \frac{1}{5} & \frac{1}{6} & \frac{1}{10} \\ \frac{1}{5} & -\frac{1}{6} & \frac{1}{10} \end{pmatrix}$

30. $\frac{1}{ad - bc}\begin{pmatrix} d & -b \\ -c & a \end{pmatrix}$ 32. $\begin{pmatrix} 0.8 & 0.0 & 0.6 \\ 0.0 & 1.0 & 0.0 \\ -0.6 & 0.0 & 0.8 \end{pmatrix}$

Exercises 15-6, p. 453

2. $x = -2$, $y = \frac{2}{5}$ 4. $x = 2$, $y = -2$ 6. $x = 3$, $y = -2$, $z = 2$
8. $x = -3$, $y = 1$, $z = \frac{1}{2}$ 10. $x = -1$, $y = 3$ 12. $x = 5$, $y = -1$ 14. $x = \frac{1}{4}$, $y = \frac{2}{3}$
16. $x = -20$, $y = 32$ 18. $x = -3$, $y = -1$, $z = 4$ 20. $x = -\frac{1}{2}$, $y = 4$, $z = -5$
22. $I_A = \frac{21}{17}$ A, $I_B = -\frac{12}{17}$ A, $I_C = -\frac{9}{17}$ A 24. \$7500, \$6200

Review Exercises for Chapter 15, p. 455

2. -14 4. 68 480 6. -36 8. 66 10. -14 12. 68 480 14. -36 16. 66 18. 30
20. 118 22. $a = 0$, $b = 6$, $c = 0$, $d = 3$, $e = -6$ 24. $x = \frac{1}{2}$, $y = -\frac{1}{2}$, $z = 1$
26. $\begin{pmatrix} 10 & -12 \\ 4 & 16 \\ 0 & -4 \end{pmatrix}$ 28. $\begin{pmatrix} -3 & 3 \\ 0 & -7 \\ 2 & -2 \\ -1 & -4 \end{pmatrix}$ 30. cannot be subtracted 32. $\begin{pmatrix} 6 & -6 \\ 0 & 14 \\ -4 & 4 \\ 2 & 8 \end{pmatrix}$ 34. $\begin{pmatrix} 29 & -8 & 37 \\ 29 & 9 & -8 \end{pmatrix}$

36. $\begin{pmatrix} 6 & -13 & 18 & -4 & 5 \\ 44 & -15 & 68 & 12 & 45 \\ 34 & -7 & 46 & -1 & 32 \end{pmatrix}$ 38. $\begin{pmatrix} 5 & 3 \\ -1 & -\frac{1}{2} \end{pmatrix}$ 40. $\begin{pmatrix} \frac{10}{437} & -\frac{3}{874} \\ \frac{21}{1748} & -\frac{25}{1748} \end{pmatrix}$ 42. $\begin{pmatrix} \frac{1}{3} & 1 & -\frac{2}{3} \\ -\frac{2}{9} & -\frac{1}{3} & \frac{4}{9} \\ \frac{5}{9} & \frac{1}{3} & -\frac{1}{9} \end{pmatrix}$

44. $\begin{pmatrix} \frac{1}{2} & -\frac{1}{2} & \frac{1}{3} \\ \frac{7}{2} & -\frac{5}{2} & 3 \\ 1 & -1 & 1 \end{pmatrix}$ 46. $x = 4$, $y = -6$ 48. $x = -7.5$, $y = 5.0$ 50. $x = 5$, $y = -\frac{1}{2}$, $z = 1$
52. $x = \frac{1}{2}$, $y = \frac{1}{3}$, $z = -\frac{1}{6}$ 54. $x = 0$, $y = 2$, $z = \frac{1}{2}$

56. $x = \frac{1}{3}$, $y = 0$, $z = -1$, $t = \frac{3}{2}$ 58. 0.5 60. $-j$ 62. $N^3 = \begin{pmatrix} 0 & 1 \\ -1 & 0 \end{pmatrix} = -N$
64. $(A + B)^2 = \begin{pmatrix} 2 & 0 \\ 0 & 2 \end{pmatrix}$, $A^2 + 2AB + B^2 = \begin{pmatrix} -2 & -6 \\ 4 & 6 \end{pmatrix}$ 66. $(B^2)^{-1} = (B^{-1})^2 = \begin{pmatrix} 3 & 4 \\ 8 & 11 \end{pmatrix}$
68. $x = 225$, $y = 120$ 70. $I_A = -\frac{12}{17}$ A, $I_B = \frac{4}{17}$ A, $I_C = \frac{8}{17}$ A
72. $v_0 = 20$ cm/s, $a = 25$ cm/s^2 74. 2200 lines/min, 2600 lines/min, 3400 lines/min
76. 2, 5, 4, 2 78. $\begin{pmatrix} 830 \\ 880 \\ 290 \end{pmatrix}$ 80. $A + 2B = \begin{pmatrix} 22 & 14 & 19 \\ 17 & 30 & 25 \end{pmatrix}$

Exercises 16-1, p. 465

2. $-2 < 3$ 4. $-8 > -18$ 6. $2 < \frac{9}{2}$ 8. $2 < 3$ 10. $x < 7$ 12. $x \geq -6$ 14. $-2 \leq x$

16. $x \leq 8$ or $x \geq 12$ 18. $0 \leq x \leq 2$ or $x > 5$ 20. $x < -4$ or $0 \leq x \leq 1$ or $x \geq 5$

22. x is less than 5 or greater than 7

24. x is greater than or equal to -1 and less than 3, or greater than 5 and less than

26.   28.   30.   32.   34.   36.

38. $T < 2°C$ or $T > 5°C$ 40. 25 000 years $< t <$ 40 000 years   42. $72.32 \text{ m} \leq \ell \leq 72.42$

44. $i = 0$ for $0 \leq t \leq 0.125$ s or 0.375 s $\leq t \leq 0.625$ s
    or 0.875 s $\leq t \leq 1$ s
  $i < 0$ for 0.125 s $< t < 0.375$ s
    or 0.625 s $< t < 0.875$ s

Exercises 16-2, p. 468

2. $x \leq 4$ 4. $x > -3$ 6. $x \geq -3$ 8. $x > \frac{4}{3}$ 10. $x \geq 2$ 12. $x > 3$

14. $x \leq 3$ 16. $x < -\frac{60}{7}$ 18. $\frac{1}{3} < x \leq \frac{7}{3}$ 20. $-\frac{3}{2} \leq x \leq \frac{3}{2}$ 22. $\frac{7}{3} < x \leq 3$ 24. $x > 1$

26. $0 \leq d < \frac{7}{3}$ m   28. $0°C \leq T < 50°C$ 30. $750 \leq y \leq 4500$ lines/min   32. $0 \leq x \leq 30\ 000$ L

Exercises 16-3, p. 474

2. $x \leq -3$ or $x \geq 0$ 4. $x < -1$ or $x > 5$ 6. $-4 \leq x \leq \frac{3}{2}$ 8. $x < -\frac{1}{3}$ or $x > \frac{1}{4}$

10. $x < -\frac{1}{3}$ or $x > -\frac{1}{3}$ 12. no values of x   14. $x \geq 0$ 16. $-2 < x < -1$ or $2 < x < 3$

18. $x < -5$ or $x > 1$ 20. $x < -3$ or $x \geq -\frac{1}{3}$ 22. $x < -2$ or $x > \frac{1}{2}$ 24. $-1 < x < 3$ or $x > 4$

26. $4 < x \leq 5$ 28. $x < 0$ or $1 < x < 2$ 30. $x \leq 0$ or $x \geq 3$
                        or $3 < x < 4$ or $x > 6$ 32. $-4 \leq x \leq -2$ or $0 \leq x < 3$

34.   36.   38.   40. $0 \leq x < 0.5$ or $2.7 < x < 3.6$

$x > 1.8$ $x > 3.2$ $x < -1.4$ $1.4 < x < 3$
                                $x > 3$

42. $0 \leq t < 10$ min   44. $3.00 < q < 4.00$ cm   46. $x > 2.9$ cm   48. 0.76 cm $< x < 2.00$ c

30

Exercises 16-4, p. 477

2. $-4 < x < 2$ 4. $x < 0$ or $x > 1$ 6. $3 \le x \le 7$ 8. $x \le -1$ or $x \ge \frac{1}{3}$ 10. $\frac{5}{2} < x < \frac{13}{2}$

12. $x < -0.5$ or $x > 7.5$ 14. $\frac{2}{9} \le x \le \frac{22}{9}$ 16. $\frac{1}{2} < x < \frac{13}{2}$ 18. $x \le -\frac{3}{2}$ or $x \ge 9$

20. $x < 15$ or $x > 66$ 22. $x < -4$ or $-2 < x < -1$ or $x > 1$
24. $-4 < x < -2$ or $-1 < x < 1$
26. $1\ 800\ 000 < p < 2\ 200\ 000$ barrels
28. $8.9\ s < t < 29.9\ s$ or $44.7\ s < t < 50.8\ s$

Exercises 16-5, p. 482

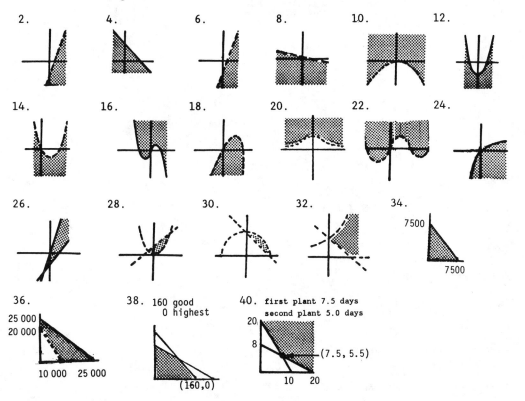

2. 4. 6. 8. 10. 12.

14. 16. 18. 20. 22. 24.

26. 28. 30. 32. 34.
7500
7500

36.
25 000
20 000
10 000 25 000

38. 160 good
0 highest
(160,0)

40. first plant 7.5 days
second plant 5.0 days
20.
8
(7.5, 5.5)
10 20

Review Exercises for Chapter 16, p. 484

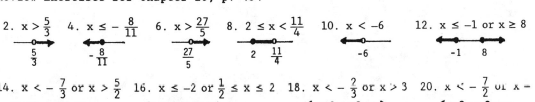

2. $x > \frac{5}{3}$ 4. $x \le -\frac{8}{11}$ 6. $x > \frac{27}{5}$ 8. $2 \le x < \frac{11}{4}$ 10. $x < -6$ 12. $x \le -1$ or $x \ge 8$

14. $x < -\frac{7}{3}$ or $x > \frac{5}{2}$ 16. $x \le -2$ or $\frac{1}{2} \le x \le 2$ 18. $x < -\frac{2}{3}$ or $x > 3$ 20. $x < -\frac{7}{2}$ or $x = 3$

31

22. $x < -4$ or $0 < x < \frac{5}{3}$ 24. $x < 2$ or $x > 6$ 26. $-3 < x < 2$ 28. $x \le 1$ or $x \ge \frac{5}{3}$

30. $\frac{5}{2} < x < \frac{13}{2}$ 32. $x \le -\frac{21}{5}$ or $x \ge \frac{19}{5}$ 34. $0 < x < \frac{1}{2}$ 36. $0 < x < 0.5$ or $1.1 < x$

38. $x \ge -5$ 40. $x < -2$ or $x \ge 1$

42.   44.   46.   48.   50.   52.

54. $85 \le x \le 105$ m   56. $0 < w < 20$ m, 30 m $< \ell < 50$ m   58. $R > 0.67$ Ω

60. 16 min $< t < 25$ min   62.   64. 50 packages at $2.00
                                             75 packages at $1.50

Exercises 17-1, p. 489

2. $\frac{3}{2}$ 4. $\frac{1}{2}$ 6. 25   8. $\frac{1}{3}$ 10. 40 000 to 1   12. 16 u   14. 2.4 A   16. 243 W
18. 10   20. 0.0131 A   22. 0.45 ha   24. 237.6 ks   26. 2.50×10^3 cm^2 28. 32 kL/h
30. 1.31 m   32. 14 clicks   34. 0.557 mA, 0.215 mA   36. $4500

Exercises 17-2, p. 495

2. $p = \frac{k}{q}$ 4. $w = kL^3$ 6. $n = \frac{k}{s^{3/2}}$ 8. $q = \frac{kr^2}{t^4}$ 10. $y = 9x$ 12. $= \frac{1}{3p^2}$ 14. $f = \frac{100}{xy}$
16. $v = 4s^2t$ 18. $\frac{75}{32}$ 20. $\frac{9}{8}$ 22. $\frac{16}{5}$ 24. 9   26. 2.09×10^5 J   28. 2.40 cm
30. 9.80 m/s^2 32. 2.31 h, k is length of trip   34. 7.65 MW   36. f/5.6
38. $P = 0.040P_0$ 40. 45.9 s   42. 20.0 cm^3 44. $\frac{1}{e^2}$ of initial value
46. 0.414d from weaker source   48. 20.2°

Review Exercises for Chapter 17, p. 497

2. $\frac{1}{20}$ 4. $\frac{5}{3}$ 6. 3.141 594   8. 311 Ω   10. 0.014 kg   12. 0.0337 L   14. 3670 kg
16. $7500, $3000   18. 3.91×10^{-5} m^3 20. 90 Mg   22. $f = \frac{40}{\ell}$ 24. $r = \frac{1}{9}uvw^2$
26. 4000 per week   28. 381 kg   30. 1.94 L   32. 10 000 J   34. 2.84 s   36. 2810 K
38. 7.3 m/s^2 40. 12.2 m/s   42. 108×10^6 km   44. 1.20 cm   46. 4   48. 1.55

2. 6, $\frac{11}{2}$, 5, $\frac{9}{2}$, 4   4. −5, −2, 1, 4, 7   6. 12   8. −34   10. $\frac{29}{3}$ 12. 86c   14. −152

16. 380k   18. d = $\frac{30}{59}$, s_{60} = 780   20. d = −3, n = 3   22. a_1 = 5, n = 10

24. d = −$\frac{3}{14}$, s_{19} = −$\frac{855}{14}$ 26. a_1 = 49c, a_{50} = −49c   28. a_7 = 5b, s_7 = 28b

30. a_1 = −$\frac{359}{5}$, d = −$\frac{6}{5}$, s_{10} = −3808   32. a_1 = 7, d = −3, s_{12} = −114   34. 10 000

36. 112   38. 272 cm, 2664 cm   40. 47.5 cm   42. 156 strokes   44. $9 375 000

46. yes   48. a_1 = 1, a_n = 2n − 1

2. 9, −6, 4, −$\frac{8}{3}$, $\frac{16}{9}$ 6. −3, −6, −12, −24, −48   6. 0.000 001   8. 8.1   10. $\frac{3}{16}$

12. −64k^5 14. $\frac{364}{3}$ 16. −180   18. a_1 = 5, s_7 = 6.249 92   20. a_1 = 8, s_7 = 5.375

22. r = 2, s_5 = 93, or r = −2, s_5 = 33   24. a_1 = −2, a_6 = 64   26. $\frac{75}{4}$(1 + √2)

28. 0.5%   30. $18 061   32. 14.7 km   34. $s_n = \frac{A(1 + i)[(1 + i)^n - 1]}{i}$ 36. 48.2%

38. 2046   40. Resulting sequence is log a + log r, log a + 2 log r, ...

2. $\frac{9}{2}$ 4. 4 + 2√2   6. 90   8. $\frac{18}{5}$ 10. $\frac{300}{13}$ 12. $\frac{1}{2}$(2 + √2)   14. $\frac{5}{9}$ 16. $\frac{7}{99}$

18. $\frac{112}{333}$ 20. $\frac{37}{45}$ 22. $\frac{548}{825}$ 24. $\frac{6152}{33\ 333}$ 26. 2.5 r   28. 800 cm²

2. x^3 − 6 x^2 + 12x − 8   4. x^8 + 12x^6 + 54x^4 + 108x^2 + 81

6. x^{5}y^5 − 5x^{4}y^{4}z + 10x^{3}y^{3}z^2 − 10x^{2}y^{2}z^3 + 5xyz^4 − z^5

8. $\frac{a^6}{x^6}$ + $\frac{6a^5}{x^4}$ + $\frac{15a^4}{x^2}$ + 20a^3 + 15a^{2}x^2 + 6ax^4 + x^6

10. b^5 + 20b^4 + 160b^3 + 640b^2 + 1280b + 1024

12. x^7 − 21x^6 + 189x^5 − 945x^4 + 2835x^3 − 5103x^2 + 5103x − 2187

14. x^8 − 24x^7 + 252x^6 − 1512x^5 + ...

16. 19 683b^9 + 118 098b^8 + 314 928b^7 + 489 888b^6 + ...

18. 2048a^{11} − 11 264$\frac{a^{10}}{x}$ + 28 160$\frac{a^9}{x^2}$ − 42 240$\frac{a^8}{x^3}$ + ...

20. 32 768x^{30} + 81 920x^{28}y + $\frac{286\ 720}{3}$x^{26}y^2 + $\frac{1\ 863\ 680}{27}$x^{24}y^3 + ...

22. 1 − $\frac{1}{3}$x + $\frac{2}{9}$x^2 − $\frac{14}{81}$x^3 + ...   24. 1 − 18x + 144x^2 − 672x^3 + ...

26. 1 − $\frac{1}{2}$x + $\frac{3}{8}$x^2 − $\frac{5}{16}$x^3 + ...   28. 2 + $\frac{1}{4}$x^2 − $\frac{1}{64}$x^4 + $\frac{1}{512}$x^6 + ...

30.(a) 35 280, (b) 8, (c) 203 212 800, (d) 7.11×10^{74}

32. $\frac{(n + 1)!}{(n - 2)!}$ = $\frac{(n + 1)(n)(n - 1)[(n - 2)!]}{(n - 2)!}$ = n^3 − n   34. 210x^{4}y^6 36. −2002a^{9}b^5

38. D^4 + 0.4D^3 + 0.06^2 + 0.004D + 0.0001

40.(a) 1 + x + x^2 + x^3, (b) 1 + x + x^2 + x^3, (c) 1 + A$\left(1 + \frac{\lambda_0^2}{\lambda^2} + \frac{\lambda_0^4}{\lambda^4} + \frac{\lambda_0^6}{\lambda^6}\right)$

2. −79   4. 1.024×10^{-4} 6. −17   8. $\frac{243}{32}$ 10. 0   12. 2059   14. 16

16. $\frac{2401}{32}$ 18. 33   20. 5   22. 10   24. $\frac{99}{4}$ (AS); $\frac{63}{4}$ (GS)   26. 64   28. $\frac{3}{2}$(3 − √3)

30. $\frac{121}{333}$ 32. $\frac{4229}{16\ 650}$ 34. s^4 + 8s^{3}t + 24s^{2}t^2 + 32sT^3 + 16t^4

36. $729n^6 - 1458n^5a + 1215n^4a^2 - 540n^3a^3 + 135n^2a^4 - 18na^5 + a^6$

38. $\frac{1}{2^{24}}x^{12} - \frac{3}{2^{20}}x^{11}y + \frac{33}{2^{19}}x^{10}y^2 - \frac{55}{2^{16}}x^9y^3 + \cdots$

40. $16\ 384s^{28} - 172\ 032\frac{s^{26}}{t} + 838\ 656\frac{s^{24}}{t^2} - 2\ 515\ 968\frac{s^{22}}{t^3} + \cdots$

42. $1 - 10x + 45x^2 - 120x^3 + \cdots$ 44. $\frac{1}{4}(1 + x + x^2 + x^3 + \cdots)$

46. $1 + \frac{1}{2}b^4 - \frac{1}{8}b^8 + \frac{1}{16}b^{12} - \cdots$ 48. $1 - x + \frac{5}{2}x^2 - \frac{15}{2}x^3 + \cdots$ 50. 25   52. 19.7%

54. 2240 m   56. 16.3 μs   58. 5 N   60. 1.0 km   62. 4.0°C   64. $1747.27

66. 10^{-20} atm   68. $-16, -4, 8, \ldots$ (AS); $-16, 8, -4, \ldots$ (GS)

Exercises 19-1, p. 528

(Note: "Answers" to trigonometric identities are intermediate steps of suggested reductions of the left member.)

2. $-0.1763 = \frac{0.1736}{-0.9848}$ 4. $1 + \left(-\frac{1}{3}\sqrt{3}\right)^2 = \left(-\frac{2}{\sqrt{3}}\right)^2$ 6. $\frac{\sin y}{\cos y}\frac{1}{\sin y} = \frac{1}{\cos y}$

8. $\frac{\frac{\sin \theta}{1}}{\cos \theta} = \frac{1}{\sin \theta}\frac{\cos \theta}{1} = \frac{\cos \theta}{\sin \theta}$ 10. $\cos x\left(\frac{\sin x}{\cos x}\right)$ 12. $\frac{\cos \theta}{\sin \theta}\frac{1}{\cos \theta} = \frac{1}{\sin \theta}$

14. $\cos^2 x(\sec^2 x)$ 16. $\sec \theta(\cos^2 \theta) = \frac{1}{\cos \theta}(\cos \theta \cos \theta)$

18. $\cos y \sec y - \cos^2 y = 1 - \cos^2 y$ 20. $\csc^2 x - \csc x \sin x = \csc^2 x - 1$

22. $\frac{1}{\cos x \sin x} - \frac{\cos x}{\sin x} = \frac{1 - \cos^2 x}{\cos x \sin x} = \frac{\sin^2 x}{\cos x \sin x} = \frac{\sin x}{\cos x}$

24. $\frac{1}{\sin x \cos x}\frac{1}{\cos x} - \frac{\sin x}{\cos x} = \frac{1 - \sin^2 x}{\sin x \cos x} = \frac{\cos^2 x}{\sin x \cos x} = \frac{\cos x}{\sin x}$

26. $\sin x \cos x\left(\frac{\sin x}{\cos x}\right) = \sin^2 x$ 28. $\sin y(1 + \cot^2 y) = \sin y(\csc^2 y) = \frac{\sin y}{\sin^2 y} = \frac{1}{\sin y}$

30. $\frac{\sin x}{\cos x} + \frac{\cos x}{\sin x} = \frac{\sin^2 x + \cos^2 x}{\cos x \sin x} = \frac{1}{\cos x \sin x} = \frac{\sin x}{\cos x \sin^2 x} = \tan x\left(\frac{1}{\sin^2 x}\right)$

32. $\tan^2 y(\sec^2 y - \tan^2 y)$ 34. $\frac{1 + \cos x}{\sin x}\frac{1 - \cos x}{1 - \cos x} = \frac{1 - \cos^2 x}{\sin x(1 - \cos x)}$

36. $1 + \frac{1}{\cot x}$ 38. $\frac{\sin \theta}{\frac{1}{\sin \theta}} + \frac{\cos \theta}{\frac{1}{\cos \theta}} = \sin^2 \theta + \cos^2 \theta$ 40. $\frac{\frac{1}{\sin \theta}}{\sin \theta} - \frac{\frac{\cos \theta}{\sin \theta}}{\frac{\sin \theta}{\cos \theta}} = \frac{1 - \cos^2 \theta}{\sin^2 \theta}$

42. $4 \sin x + \frac{\sin x}{\cos x} = \sin x\left(4 + \frac{1}{\cos x}\right)$ 44. $\frac{1}{\sin x} + \frac{1}{\cos x} - \frac{1}{\cos x} = \frac{1}{\sin x}$

46. $\frac{1}{\sin^2 \theta} - \frac{\cos \theta}{\sin^2 \theta} = \frac{1 - \cos \theta}{1 - \cos^2 \theta}$ 48. $\frac{\frac{1}{\cos x} - \cos x}{\frac{\sin x}{\cos x}} = \frac{1 - \cos^2 x}{\cos x}\frac{\cos x}{\sin x}$

50. $(2 \sin^2 x - 1)(\sin^2 x - 1)$

52. $\frac{\sin 5y}{2}\left[\frac{\sin^2 5y + (1 - \cos 5y)^2}{(1 - \cos 5y)\sin 5y}\right] = \frac{\sin^2 5y + 1 - 2 \cos 5y + \cos^2 5y}{2(1 - \cos 5y)} = \frac{2(1 - \cos 5y)}{2(1 - \cos 5y)}$

54. $\frac{\cot 2y(\sec 2y + \tan 2y) - (\cos 2y)(\sec 2y - \tan 2y)}{\sec^2 2y - \tan^2 2y}$

$= \frac{\cot 2y \sec 2y + 1 - 1 + \cos 2y \tan 2y}{1}$

34

56. Infinite GS: $\dfrac{1}{1 + \tan^2 x} - \dfrac{1}{\sec^2 x}$

58. $r^2\sin^2\theta + R^2 - r^2 = R^2 + r^2(\sin^2\theta - 1) = R^2 + r^2(-\cos^2\theta)$

60. $\sqrt{\dfrac{(1 + \cos\theta)(1 + \cos\theta)}{(1 - \cos\theta)(1 + \cos\theta)}}\sin\theta = \sqrt{\dfrac{(1 + \cos\theta)^2}{1 - \cos^2\theta}}\sin\theta$

62. $\sqrt{9 - 9\sin^2\theta} = 3\sqrt{1 - \sin^2\theta} = 3\cos\theta$

64. $\sqrt{16\sec^2\theta - 16} = 4\sqrt{\sec^2\theta - 1} = 4\tan\theta$

Exercises 19-2, p. 533

2. $\cos 75° = \cos 30°\cos 45° - \sin 30°\sin 45° = \left(\tfrac{1}{2}\sqrt{3}\right)\left(\tfrac{1}{2}\sqrt{2}\right) - \tfrac{1}{2}\left(\tfrac{1}{2}\sqrt{2}\right) = \dfrac{\sqrt{6} - \sqrt{2}}{4} = 0.2588$

4. $\sin 15° = \sin 45°\cos 30° - \cos 45°\sin 30° = \left(\tfrac{1}{2}\sqrt{2}\right)\left(\tfrac{1}{2}\sqrt{3}\right) - \left(\tfrac{1}{2}\sqrt{2}\right)\left(\tfrac{1}{2}\right) = \dfrac{\sqrt{6} - \sqrt{2}}{4} = 0.2588$

6. $-\dfrac{16}{65}$ 8. $-\dfrac{63}{65}$ 10. $\sin 2x$ 12. $\cos 2x$ 14. $\sin(2x + 1)$ 16. $\cos 2\pi = 1$

18. -1 20. 0 22. $\cos 180°\cos x + \sin 180°\sin x = (-1)\cos x + (0)\sin x$

24. $\sin(0 - x) = \sin 0 \cos x - \cos 0 \sin x = (0)\cos x - (1)\sin x$

26. $\sin(90° + x) = \sin 90°\cos x + \cos 90°\sin x = (1)\cos x + (0)\sin x$

28. $\cos\left(\tfrac{3}{2}\pi + x\right) = \cos\tfrac{3}{2}\pi \cos x - \sin\tfrac{3}{2}\pi \sin x = (0)\cos x - (-1)\sin x$

30. $\sin(120° - x) = \sin 120°\cos x - \cos 120°\sin x$
$$= \tfrac{1}{2}\sqrt{3}\cos x + \tfrac{1}{2}\sin x = \tfrac{1}{2}(\sqrt{3}\cos x + \sin x)$$

32. $\cos\left(\tfrac{1}{3}\pi + x\right) = \cos\tfrac{1}{3}\pi \cos x - \sin\tfrac{1}{3}\pi \sin x = \tfrac{1}{2}\cos x - \tfrac{1}{2}\sqrt{3}\sin x$

34. $(\cos x \cos y - \sin x \sin y)(\cos x \cos y + \sin x \sin y)$
$$= \cos^2 x \cos^2 y - \sin^2 x \sin^2 y = \cos^2 x(1 - \sin^2 y) - \sin^2 x \sin^2 y$$

36. $\cos(x - y) + \sin(x + y) = \cos x \cos y + \sin x \sin y + \sin x \cos y + \cos x \sin y$
$$= \cos x(\cos y + \sin y) + \sin x(\sin y + \cos y)$$

38., 40., 42., 44., Use indicated method

46. $w = T\cos\alpha - \left(T\dfrac{\sin\alpha}{\cos\theta}\right)\sin\theta = \dfrac{T\cos\alpha\cos\theta - T\sin\alpha\sin\theta}{\cos\theta}$

48. $E_2 = E_1\dfrac{\tan r - \tan i}{\tan r + \tan i} = E_1\dfrac{\dfrac{\sin r}{\cos r} - \dfrac{\sin i}{\cos i}}{\dfrac{\sin r}{\cos r} + \dfrac{\sin i}{\cos i}} = E_1\dfrac{\sin r \cos i - \cos r \sin i}{\sin r \cos i + \cos r \sin i}$

Exercises 19-3, p. 538

2. $\sin 120° = \sin 2(60°) = 2\sin 60°\cos 60° = 2\left(\tfrac{1}{2}\sqrt{3}\right)\left(\tfrac{1}{2}\right) = \tfrac{1}{2}\sqrt{3}$

4. $\cos 60° = \cos 2(30°) = \cos^2 30° - \sin^2 30° = \left(\tfrac{1}{2}\sqrt{3}\right)^2 - \left(\tfrac{1}{2}\right)^2 = \tfrac{1}{2}$

6. $\sin 84° = 2\sin 42°\cos 42° = 0.994\,521\,9$

8. $\cos 276° = \cos^2 138° - \sin^2 138° = 0.104\,528\,5$

10. $-\frac{119}{169}$ 12. $\frac{336}{625}$ 14. $\sin^2 2x$ 16. $-\cos 8x$ 18. $\sin x$ 20. $\frac{1}{2}\sin 6x$

22. $(1 - \sin^2\alpha) - \sin^2\alpha$ 24. $2\frac{\sin x}{\cos x}\cos x$

26. $\sin^2 x + 2\sin x \cos x + \cos^2 x = (\sin^2 x + \cos^2 x) + 2\sin x \cos x$

28. $\dfrac{2\sin^2\theta + \cos^2\theta - \sin^2\theta}{\sin^2\theta}$ 30. $\dfrac{2\sin\alpha}{\cos\alpha}\dfrac{1}{\sec^2\alpha} = 2\sin\alpha\cos\alpha$

32. $1 - (1 - 2\sin^2\theta) = \dfrac{2}{\csc^2\theta}$ 34. $2\sin x + 2\sin x \cos x = \dfrac{2\sin x(1 + \cos x)(1 - \cos}{1 - \cos x}$

36. $\dfrac{\sin 3x \cos x + \cos 3x \sin x}{\sin x \cos x} = \dfrac{\sin 4x}{\frac{1}{2}\sin 2x} = \dfrac{2\sin 2x \cos 2x}{\frac{1}{2}\sin 2x}$

38. $\cos(2x + x) = \cos 2x \cos x - \sin 2x \sin x = (\cos^2 x - \sin^2 x)\cos x - (2\sin x \cos x)\sin$

40. $x = 1 - 2y^2$ 42. $\sin\frac{\pi}{2}\cos 2\theta - \cos\frac{\pi}{2}\sin 2\theta = \cos 2\theta$

44. $s = a\left[\frac{1}{2}(1 + \cos 2\theta)\right] + b\left[\frac{1}{2}(1 - \cos 2\theta)\right] - t(2\sin\theta\cos\theta)$

Exercises 19-4, p. 542

2. $\sin 22.5° = \sin\frac{1}{2}(45°) = \sqrt{\dfrac{1 - \cos 45°}{2}} = \sqrt{\dfrac{0.2929}{2}} = 0.3827$

4. $\cos 112.5° = \cos\frac{1}{2}(225°) = -\sqrt{\dfrac{1 + \cos 225°}{2}} = -\sqrt{\dfrac{0.2929}{2}} = -0.3827$

6. $\cos 49° = 0.656\,059\,0$ 8. $2\sin 164° = 0.551\,274\,7$ 10. $2\cos 4\beta$ 12. $2\sin 8x$

14. $-\frac{1}{5}\sqrt{5}$ 16. $\frac{3}{34}\sqrt{34}$ 18. $\pm\sqrt{\dfrac{2\sec\alpha}{\sec\alpha + 1}}$ 20. $\dfrac{1 + \cos\alpha}{\sin\alpha} = \dfrac{\sin\alpha}{1 - \cos\alpha}$

22. $2\sqrt{\dfrac{2(1 + \cos x)(1 + \cos x)}{2(2)(1 + \cos x)}} = \dfrac{2(1 + \cos x)}{2}\sqrt{\dfrac{2}{1 + \cos x}}$ 24. $2\left(\dfrac{1 + \cos\theta}{2}\right)\sec\theta$

26. $\dfrac{1 + \cos x}{2}\left(1 + \dfrac{\sin^2 x}{(1 + \cos x)^2}\right) = \dfrac{1 + 2\cos x + \cos^2 x + \sin^2 x}{2(1 + \cos x)}$

28. $\tan\frac{1}{2}\alpha = \dfrac{\sin\frac{1}{2}\alpha}{\cos\frac{1}{2}\alpha} = \dfrac{2\cos\frac{1}{2}\alpha\sin\frac{1}{2}\alpha}{2\cos^2\frac{1}{2}\alpha}$ 30. $2\left(\dfrac{1 - \cos\alpha}{2} - \dfrac{1 - \cos\theta}{2}\right)$

32. $\cos\theta = \dfrac{\ell - x}{\ell}$; $x = \ell(1 - \cos\theta) = \ell[1 - (1 - 2\sin^2\frac{1}{2}\theta)]$

Exercises 19-5, p. 546

2. $\frac{7}{6}\pi, \frac{11}{6}\pi$,   4. $\frac{2}{3}\pi, \frac{4}{3}\pi$ 6. $\frac{1}{4}\pi, \frac{5}{4}\pi$ 8. $1.9823, 4.3009$ 10. $\frac{1}{2}\pi, \frac{3}{2}\pi$

12. $\frac{1}{6}\pi, \frac{5}{6}\pi, \frac{7}{6}\pi, \frac{11}{6}\pi$ 14. $0.7227, \frac{1}{2}\pi, \frac{3}{2}\pi, 5.560$

16. $0, \frac{1}{6}\pi, \frac{1}{2}\pi, \frac{5}{6}\pi, \pi, \frac{7}{6}\pi, \frac{3}{2}\pi, \frac{11}{6}\pi$ 18. $\frac{1}{2}\pi, \frac{3}{2}\pi$ 20. $0, \frac{2}{3}\pi$

22. $0, \frac{1}{6}\pi, \frac{5}{6}\pi, \pi, \frac{7}{6}\pi, \frac{11}{6}\pi$ 24. $1.107, 1.249, 4.249, 4.391$

26. $0.9553, 2.186, 4.097, 5.328$ 28. 0

30. $\frac{1}{24}\pi, \frac{1}{8}\pi, \frac{5}{24}\pi, \frac{13}{24}\pi, \frac{5}{8}\pi, \frac{17}{24}\pi, \frac{25}{24}\pi, \frac{9}{8}\pi, \frac{29}{24}\pi, \frac{37}{24}\pi, \frac{13}{8}\pi, \frac{41}{24}\pi$

32. $\frac{1}{8}\pi, \frac{3}{8}\pi, \frac{5}{8}\pi, \frac{7}{8}\pi, \frac{9}{8}\pi, \frac{11}{8}\pi, \frac{13}{8}\pi, \frac{15}{8}\pi$ 34. $0.9553 = 54.7°$ 36. 360 m

38. $2.4189, 3.8643$ 40. $1.1314, 2.0102, 3.2258, 4.1046, 5.3202, 6.1990$

42. $2.1, -2.1$ 44. 0.95

2. y is the angle whose secant is x.   4. y is the angle whose cosecant is 4x.

6. y is three times the angle whose tangent is x.

8. y is four times the angle whose sine is 3x.   10. $\frac{1}{2}\pi$ 12. $\frac{1}{2}\pi$ 14. $-\frac{1}{6}\pi$

16. $\frac{1}{6}\pi$ 18. $\frac{1}{4}\pi$ 20. $\frac{5}{6}\pi$ 22. $\frac{1}{4}\pi$ 24. $\frac{5}{6}\pi$ 26. $-\frac{1}{2}\pi$ 28. 1   30. -2 32. 0.8000

34. 2.286   36. 0.2773   38. 0.2135   40. -1.093 42. 0.632 19   44. -0.3167

46. $x = \pi + \text{Arccos } y$ 48. $x = 6 \sin\frac{1}{2}y$ 50. $x = \frac{1}{8}\text{Arccsc}(5 - 4y)$ 52. $x = \frac{1}{3}\cot(2y + 5)$

54. $\sqrt{1 - x^2}$ 56. x   58. $\frac{2x}{\sqrt{1 - 4x^2}}$ 60. $\frac{1 - x^2}{1 + x^2}$ 62. $\theta = \text{Arctan } \mu$ 64. $d = A \cos 2\pi ft$

66. $\sin\left(\text{Arctan }\frac{1}{3} + \text{Arctan }\frac{1}{2}\right) = \frac{1}{\sqrt{10}}\frac{2}{\sqrt{5}} + \frac{3}{\sqrt{10}}\frac{1}{\sqrt{5}} = \frac{1}{\sqrt{2}} = \sin\frac{1}{4}\pi$ 68. $\frac{1}{3}\pi + \frac{1}{6}\pi = \frac{1}{2}\pi$

70. $A = \text{Arcsin}\frac{a \sin B}{b}$ 72. $L = 12 + 12 + 8\pi + 3\pi + (16 - 6)\text{Arcsin}\frac{5}{13}$

Review Exercises for Chapter 19, p. 555

2. $\cos(90° - 60°) = \cos 90°\cos 60° + \sin 90°\sin 60° = 0\left(\frac{1}{2}\right) + (1)\left(\frac{1}{2}\sqrt{3}\right) = \frac{1}{2}\sqrt{3}$

4. $\cos(180° + 45°) = \cos 180°\cos45° - \sin 180°\sin 45° = (-1)\left(\frac{1}{2}\sqrt{2}\right) - (0)\left(\frac{1}{2}\sqrt{2}\right) = -\frac{1}{2}\sqrt{2}$

6. $\sin 2(90°) = 2 \sin 90°\cos 90° = 2(1)(0) = 0$

8. $\cos \frac{1}{2}(90°) = \sqrt{\frac{1 + \cos 90°}{2}} = \sqrt{\frac{1 + 0}{2}} = \frac{1}{2}\sqrt{2}$ 10. $\cos 296° = 0.438\ 371\ 1$

12. $\cos(-69°) = 0.358\ 367\ 9$ 14. $\cos 218° = -0.788\ 010\ 8$ 16. $\sin 83° = 0.992\ 546\ 2$

18. $\cos 4x$ 20. $5 \sin 10x$ 22. $\cos 4x$ 24. $8 \sin 2x$ 26. $\frac{1}{4}\pi$ 28. -0.5585 30. $\frac{1}{2}$

32. π 34. $\cos\theta\left(\frac{1}{\sin\theta}\right) = \frac{\cos\theta}{\sin\theta}$ 36. $\cos y \sec y - \cos^2 y = 1 - \cos^2 y$

38. $\sin\theta\left(\frac{1}{\cos\theta}\right)\left(\frac{1}{\sin\theta}\right)\cos\theta$ 40. $\sin x + \frac{\cos x}{\frac{\cos x}{\sin x}} = \sin x + \sin x$

42. (Change right side.) $\dfrac{1 - \frac{\sin^2 y}{\cos^2 y}}{1 + \frac{\sin^2 y}{\cos^2 y}} = \dfrac{\cos^2 y - \sin^2 y}{\cos^2 y + \sin^2 y}$

44. $\sin x(\csc^2 x - 1) = (\sin x \csc x)\csc x - \sin x$ 46. $\dfrac{1 + 2 \cos^2\theta - 1}{\cos^2\theta}$

48. $\dfrac{2 \sin\theta \cos\theta}{2 \cos\theta}$ 50. $\sqrt{\dfrac{1 - \cos x}{2}} = \sqrt{\dfrac{1 - \frac{1}{\sec x}}{2}} = \sqrt{\dfrac{\frac{\sec x - 1}{2 \sec x}}{\frac{\sec x - 1}{2 \sec x}}}\dfrac{2 \sec x}{2 \sec x}$

$= \sqrt{\dfrac{\sec x - 1}{2 \sec x}}\sqrt{\dfrac{2 \sec x}{\sec x - 1}} = \sqrt{\dfrac{\sec x - 1}{2 \sec x}}\sqrt{\dfrac{2}{1 - \cos x}}$

52. $\dfrac{\frac{\cos\theta - \sin\theta}{\sin\theta}}{\frac{\cos\theta + \sin\theta}{\sin\theta}}$ 54. $\sin(3y - 2y)$ 56. $\dfrac{1}{\sin 2x} + \dfrac{\cos 2x}{\sin 2x} = \dfrac{2 \cos^2 x}{2 \sin x \cos x}$

58. (Change right side.) $1 - \sin\frac{x}{2} - 2 \sin^2\frac{x}{2} = \left(1 - 2 \sin^2\frac{x}{2}\right) - \sin\frac{x}{2}$

60. $\dfrac{1}{\cos\frac{x}{2}} + \dfrac{1}{\sin\frac{x}{2}} = \dfrac{\sin\frac{x}{2} + \cos\frac{x}{2}}{\cos\frac{x}{2}\sin\frac{x}{2}}$ 62. $x = \frac{1}{2}\pi + \text{Arctan }\frac{1}{2}(y - 2)$ 64. $x = \frac{1}{4}\sec(2y + 2)$

66. 0.1674, 2.9741   68. $\frac{1}{4}\pi, \frac{5}{4}\pi$ 70. $\frac{7}{12}\pi, \frac{11}{12}\pi, \frac{19}{12}\pi, \frac{23}{12}\pi$ 72. $\frac{1}{6}\pi, \frac{5}{6}\pi, \frac{3}{2}\pi$

74. $\frac{1}{4}\pi, \frac{3}{4}\pi, \frac{5}{4}\pi, \frac{7}{4}\pi$ 76. 0, $\frac{1}{2}\pi$ 78. $\frac{\sqrt{x^2 - 1}}{x}$ 80. $-\frac{1}{\sqrt{1 + x^2}}$

82. $\sqrt{4\sec^2\theta - 4} = 2\sqrt{\sec^2\theta - 1} = 2\tan\theta$ 84. $\dfrac{\sqrt{1 - \cos^2\theta}}{\cos\theta} = \dfrac{\sin\theta}{\cos\theta} = \tan\theta$ 86. $\dfrac{\cos 2\alpha}{2\cos^2\alpha}$

88. $\dfrac{\sin x \cos\frac{1}{2}x + \cos x \sin\frac{1}{2}x}{\sin\frac{1}{2}x}\!-\!\sin x = \dfrac{2\sin\frac{1}{2}x \cos^2\frac{1}{2}x + \cos x \sin\frac{1}{2}x}{\sin\frac{1}{2}x}\!-\!\sin x$

$= \left(2\cos^2\frac{1}{2}x + \cos x\right)\sin x = (1 + 2\cos x)\sin x$

90. $t = \dfrac{1}{\omega}(\alpha - \phi + \text{Arcsin } 0.4p)$

92. $(\tan^2 A + 1) - \sin^2 B \tan^2 A = \sec^2 C; \ \tan^2 A(1 - \sin^2 B) = \sec^2 C - 1$

94. 91.7 m   96. 0.015 s

Exercises 20-1, p. 563

2. $7\sqrt{2}$ 4. $\sqrt{34}$ 6. 52   8. 7   10. 16.9   12. 1   14. $\dfrac{3}{5}$ 16. $-\dfrac{5}{12}$

18. undefined   20. −0.71   22. $\sqrt{3}$ 24. −1   26. 39.5°   28. 125°

30. perpendicular   32. parallel   34. $\pm 2\sqrt{5}$ 36. $-4 \pm 3\sqrt{3}$ 38. $m_1 = \dfrac{1}{2}, \ m_2 = -2$

40. All sides are $\sqrt{29}$, slopes are $\dfrac{2}{5}$ and $-\dfrac{5}{2}$ 42. 29   44. 36

46. (−7, −1)   48. (−0.8, 1.3)

Exercises 20-2, p. 569

2. $2x + y + 5 = 0$ 4. $2x + y + 1 = 0$ 6. $\sqrt{3}x + y + 2 = 0$ 8. $x = -4$ 10. $y = -5$

12. $2x - y + 6 = 0$ 14. $x - 4y + 12 = 0$ 16. $x + 4y - 9 = 0$ 18. $x + 7y - 18 = 0$

20. $3x - y + 7 = 0$ 22.   24.

26. $y = -2x + \dfrac{5}{2}; \ m = -2, \ \left(0, \dfrac{5}{2}\right)$

28. $y = 2x - \dfrac{4}{3}; \ m = 2, \ \left(0, -\dfrac{4}{3}\right)$

30. 8   32. −9   34. $m_1 = m_2 =$

36. $m_1 = 4; \ m_2 = -\dfrac{1}{4}$

38. $8x - 32y - 3 = 0$

40. $x + y - 1 = 0$

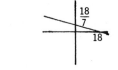

42. $V = 6.00 - 180i$

44. $2x + 3y = 200$ 46. $T = \dfrac{4}{3}x + 3$

48. $p = 4w + 20$

50. $y = 3.38x$

52. $w = 30 - 2t$ 54.   56.   58. $m = 0.3010$ 60. $v = 40(0.952)^t$

38

Exercises 20-3, p. 574

2. $(3, -4)$, $r = 7$ 4. $(0, 6)$, $r = 8$ 6. $x^2 + y^2 = 1$

8. $x^2 + (y - 2)^2 = 4$ or $x^2 + y^2 - 4y = 0$

10. $(x + 3)^2 + (y + 5)^2 = 12$ or $x^2 + y^2 + 6x + 10y + 22 = 0$

12. $\left(x - \frac{3}{2}\right)^2 + (y + 2)^2 = \frac{25}{4}$ or

$4x^2 + 4y^2 - 12x + 16y = 0$

14. $(x + 1)^2 + (y - 4)^2 = 2$ or $x^2 + y^2 + 2x - 8y + 15 = 0$

16. $(x - 2)^2 + (y + 4)^2 = 4$ or $x^2 + y^2 - 4x + 8y + 16 = 0$

18. $x^2 + y^2 + 8x - 8y + 16 = 0$ 20. $x^2 + y^2 - 8y - 9 = 0$

22. $(2, -3)$ 24. $(-4, -3)$ 26. $(0, 0)$ 28. $(2, 3)$ 30. $(-4, -3)$ 32. $(2, 0)$

$r = 7$ $r = \frac{5}{2}\sqrt{2}$ $r = 3$ $r = 5$ $r = 5$ $r = \frac{2}{3}\sqrt{6}$

34. symmetric to x-axis   36. not symmetric to axes or origin   38. $(1, 0)$, $(2, 1)$

40. $x^2 + y^2 = 4$, circle   42. $V_H^2 + V_V^2 = 1.24 \times 10^8$ 44. 4.57 m^2

Exercises 20-4, p. 579

2. $F(4, 0)$, $x = -4$ 4. $F(-4, 0)$, $x = 4$ 6. $F\left(0, \frac{5}{2}\right)$, $y = -\frac{5}{2}$ 8. $F(0, -3)$, $y = 3$

10. $F\left(0, \frac{7}{2}\right)$, $y = -\frac{7}{2}$ 12. $F\left(\frac{1}{12}, 0\right)$, $x = -\frac{1}{12}$ 14. $y^2 = -8x$ 16. $y^2 = -12x$

18. $x^2 = -2y$ 20. $y^2 = \frac{1}{2}x$

22. $y^2 + 2x + 8y + 13 = 0$ 24. $y^2 + 8y + 20x + 56 = 0$ 26. $x^2 + y^2 - 2y = 0$

28. $y^2 = 1.2x$

30. $y = \frac{g}{2v_0^2}x^2$

32. 10.5 m

34.   36.

Exercises 20-5, p. 585

2. $V(10, 0)$, $V(-10, 0)$ 4. $V(0, 9)$, $V(0, -9)$ 6. $V(12, 0)$, $V(-12, 0)$

   $F(6, 0)$, $F(-6, 0)$ $F(0, 4\sqrt{2})$, $F(0, -4\sqrt{2})$ $F(2\sqrt{35}, 0)$, $F(-2\sqrt{35}, 0)$

8. V(0, 5), V(0, −5)   10. V($\sqrt{3}$, 0), V(−$\sqrt{3}$, 0)   12. V$\left(0, \frac{3}{2}\right)$, V$\left(0, -\frac{3}{2}\right)$

F(0, 2$\sqrt{6}$), F(0, −2$\sqrt{6}$)   F(1, 0), F(−1, 0)   F$\left(0, \frac{1}{2}\sqrt{5}\right)$, F$\left(0, -\frac{1}{2}\sqrt{5}\right)$

     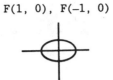

14. $25x^2 + 16y^2 = 400$ 16. $4x^2 + 13y^2 = 52$ 18. $3x^2 + y^2 = 6$ 20. $2x^2 + 3y^2 = 20$

22. $169x^2 + 144y^2 + 1014x - 864y - 21{,}519 = 0$ 24. $4x^2 + 3y^2 + 16x - 18y + 31 = 0$

26. $5(-x)^2 + y^2 - 3y - 7 = 5x^2 + y^2 - 3y - 7 = 0$ 28. 0.24

30. $i_1^2 + 4i_2^2 = 32$ 32. $x^2 + 49y^2 = 1.96$ 34. 4.08 m   36. 41.1 mm^2

Exercises 20-6, p. 592

2. V(4, 0), V(−4, 0)   4. V(0, $\sqrt{2}$), V(0, −$\sqrt{2}$)   6. V(9, 0), V(−9, 0)

F(2$\sqrt{5}$, 0), F(−2$\sqrt{5}$, 0)   F(0, 2), F(0, −2)   F(3$\sqrt{10}$, 0), F(−3$\sqrt{10}$, 0)

8. V(0, $\sqrt{2}$), V(0, −$\sqrt{2}$)   10. V(1, 0), V(−1, 0)   12. V(0, 5), V(0, −5)

F(0, $\sqrt{5}$), F(0, −$\sqrt{5}$)   F($\sqrt{10}$, 0), F(−$\sqrt{10}$, 0)   F$\left(0, \frac{5}{3}\sqrt{10}\right)$, F$\left(0, -\frac{5}{3}\sqrt{10}\right)$

14. $2y^2 - x^2 = 2$ 16. $15x^2 - y^2 = 60$ 18. $x^2 - 16y^2 = 16$ 20. $3y^2 - 4x^2 = 8$

22.   24.   26. $4y^2 - 9x^2 - 8y - 36x - 68 = 0$

28. $9x^2 - 16y^2 - 18x - 32y + 137 = 0$

30. $c = \sqrt{\frac{5}{3}} = 1.29$ 32. $\pi R^2 - 2\pi r^2 = 24$

34.   36. $3x^2 - y^2 = 3$

Exercises 20-7, p. 596

2. ellipse, (−4, 1)   4. parabola, (2, −5)   6. hyperbola, (−2, 4)   8. ellipse, (0,

10. $(x - 2)^2 = -16(y + 1)$ or $x^2 - 4x + 16y + 20 = 0$

12. $(y - 4)^2 = -8(x - 4)$ or $y^2 - 8y + 8x - 16 = 0$

14. $\dfrac{x^2}{169} + \dfrac{(y-3)^2}{25} = 1$ or $25x^2 + 169y^2 - 1014y - 2704 = 0$

16. $\dfrac{4(y-4)^2}{169} + \dfrac{4(x-1)^2}{25} = 1$ or $676x^2 + 100y^2 - 1352x - 800y - 1949 = 0$

18. no hyperbola (b = c)

20. $\dfrac{(y+4)^2}{16} - \dfrac{(x-1)^2}{9} = 1$ or $9y^2 - 16x^2 + 32x + 72y - 16 = 0$

22. parabola, (−5,1) 24. ellipse, (−2,4) 26. hyperbola, (−2,1) 28. parabola, (−2,4)

30. $x^2 + 2y^2 + 4x - 14 = 0$ 32. Use Eqs. (20-27)

34. 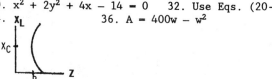 36. $A = 400w - w^2$

Exercises 20-8, p. 600

2. parabola   4. circle   6. hyperbola   8. ellipse   10. ellipse
12. parabola   14. circle   16. parabola   18. hyperbola   20. circle
22. ellipse   24. circle   26. hyperbola   28. ellipse

C(3, 0)   C(−1, 0), $r = \dfrac{3}{2}$ C(−2, −2)   C(0, 1)

V($3\pm2\sqrt{2}$, 0)   V(0, −2)   V(0, 4)
                                               V(−4, −2)   V(0, −2)

30.(a) ellipse, (b) hyperbola   32. no real locus   34. parabola   36. hyperbola

Exercises 20-9, p. 604

2.   4.   6.   8.   10.   12.

14. $\left(\sqrt{2}, \dfrac{5}{4}\pi\right)$ 16. (6.40, 2.47)   18. (4, 0)   20. (−0.54, −0.84)   22. $r = 2\csc\theta$

24. $r = 4\sin\theta$ 26. $r^2 = a^2\sec 2\theta$ 28. $r = \tan\theta\sec\theta$ 30. $x^2 + y^2 = 4x$ 32. $y = -2$

34. $x^4 + y^4 + 2x^2y^2 + 2x^2y + 2y^3 - x^2 = 0$ 36. $(x^2 + y^2)^2 = 16(x^2 - y^2)$

38. $x^2 + y^2 = 1.69$, $r = 1.30$ 40. $0.98x^2 + y^2 + 2130x - 5.78\times10^7 = 0$

Exercises 20-10, p. 607

2.   4.   6.   8.   10.   12.

41

14. 16. 18. 20. 22.

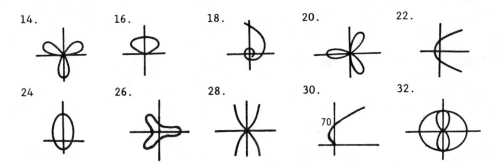

Review Exercises for Chapter 20, p. 609

2. $8x - y + 13 = 0$ 4. $2x - 5y - 4 = 0$ 6. $x^2 + y^2 - 10x - 2y + 22 = 0$

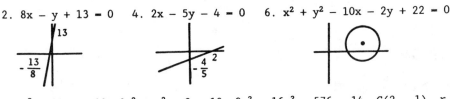

8. $x^2 = 20y$ 10. $2x^2 + y^2 = 9$ 12. $9y^2 - 16x^2 = 576$ 14. $C(2, -1)$, $r = 5$

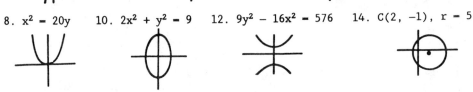

16. $F(6, 0)$, $x = -6$ 18. $V(0, 3)$, $V(0 - 3)$ 20. $V(5, 0)$, $V(-5,0)$
         $F(0, \sqrt{11})$, $F(0, -\sqrt{11})$ $F(\sqrt{23}, 0)$, $F(-\sqrt{23}, 0)$

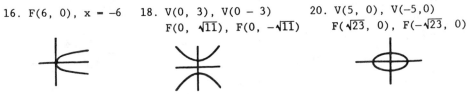

22. $V(5, -2)$ 24. $C(-2,1)$ 26. 28. 30. 32.
  $x = 4$

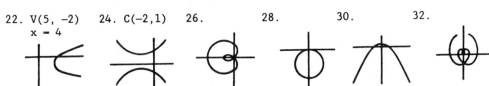

34. $r^2\sin 2\theta = 1$ 36. $r^2 + 6r \sin \theta - 7 = 0$ 38. $(x^2 + y^2)^3 = y^2$ 40. $x^2 - 4y - 4 = 0$

42. 1 44. 2 46. $(0, 1)$, $(0, -1)$ 48. Point: $\left(\frac{60}{7}, \frac{31}{21}\right)$ 50. $x^2 + 4y - 12 = 0$

52. $v = 6.0t + 5.0$ 54. 56. $A = 2t + 4$ 58. $x^2 + y^2 = 4$; $(x - 8)^2 + y^2 = 3$

60. 62. 64. 66. 2.30×10^4 m^2

              68. 18 cm, 8 cm

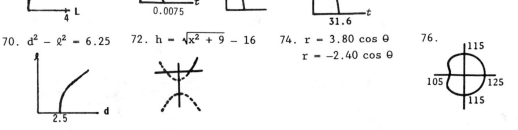

70. $d^2 - \ell^2 = 6.25$ 72. $h = \sqrt{x^2 + 9} - 16$ 74. $r = 3.80 \cos \theta$ 76.
                 $r = -2.40 \cos \theta$

Exercises 21-1, p. 617

2.

No.	23	24	25	26	27	28
Freq.	2	1	3	2	1	2

4.

No.	103	104	105	106	107	108	109	110	111	112	113
Freq.	1	3	2	3	2	5	3	1	1	0	1

6.

Int.	22-24	25-27	28-30
Freq.	3	6	2

8.

Int.	101-105	106-110	111-115
Freq.	6	14	2

10. 　　12. 　　14.　　16.　　18.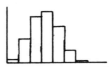

20.

Int.	17-19	20-22	23-25
Freq.	4	9	2

22.　　24.

26.　　28.　　30.　　32. (50 stocks)

Exercises 21-2, p. 622

2. 25   4. 107.5   6. 25.5   8. 107.2   10. 25   12. 108   14. 20.9   16. 2.25 s
18. 2.25 s   20. 56.5 m   22. 4.265 mR   24. 31.2 h   26. 30 h   28. 0.0059 mm
30. $277   32. 900 kW·h, 900 kW·h   34. 25.5   36. 108

Exercises 21-3, p. 628

2. 1.67   4. 2.44   6. 1.67   8. 2.44   10. 0.174 mR   12. $60   14. 3.3 m
16. 147 kW·h   18. 64%   20. 68%   22. 75%   24. 57%

Exercises 21-4, p. 632

2. $y = 10.2x - 2.3$ 4. $y = -1.11x + 15.5$ 6. $s = -0.043F + 3.69$ 8. $T = 0.32t + 20.4$

10. L = −0.86t + 8.33   12. P = 0.494T + 133   14. 0.995   16. 0.646

   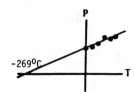

Exercises 21-5, p. 637

2. $y = 3.5\sqrt{x} + 1.3$ 4. $y = 0.99(10^x) + 5.05$ 6. $y = 0.000383x^2 + 0.50$

8. $f = 0.086\sqrt{A} - 0.040$ 10. $f = \dfrac{488}{\sqrt{L}} + 6$ 12. $i = 10.0 \sin 377t$

                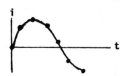

Review Exercises for Chapter 21, p. 639

2. 3.5   4. $\dfrac{2.0\text{-}2.9 \mid 3.0\text{-}3.9 \mid 4.0\text{-}4.9}{3 \quad\mid\quad 4 \quad\mid\quad 4}$ 6. 107.5   8. 107.9   10.

12. $\dfrac{101\text{-}105 \mid 106\text{-}110 \mid 111\text{-}115}{6 \quad\mid\quad 8 \quad\mid\quad 6}$ 14. 0.26 Pa·s   16.

18. 0.26 Pa·s   20. 700 W   22.   24.   26. 4.2   28.

30. n = 0.45t + 7.9   32. y = 1.17x + 9.99   34. $y = \dfrac{3.40\times10^4}{x}$

36. $\bar{y} = \dfrac{\overline{xy} - \bar{x}\,\bar{y}}{s_x^2} + \dfrac{\overline{x^2}\,\bar{y} - \overline{xy}\,\bar{x}}{s_x^2} = \dfrac{\overline{x^2}\,\bar{y} - \bar{x}^2\bar{y}}{s_x^2} = \dfrac{\bar{y}(\overline{x^2} - \bar{x}^2)}{s_x^2} = \bar{y}$

44

Exercises 22-1, p. 649

2. cont. all x   4. not cont. x = 0 and x = 1, div. by zero
6. cont. −2 < x < 0 or x > 0; not cont. x = 0, div. by zero
8. cont. all x   10. not cont. x = 2, function not defined
12. cont. −2 < x < 2; f(−2) not defined; small change at x = 2
14. cont. all x   16. not cont. x = −2, f(x) not defined

18.

x	0.4000	0.4900	0.4990	0.4999	0.5001	0.5010	0.5100	0.6000
f(x)	−1.2800	−1.4798	−1.4980	−1.4998	−1.5002	−1.5020	−1.5198	−1.6800

$\lim\limits_{x \to 0.5} f(x) = -1.5$

20.

x	−3.100	−3.010	−3.001	−2.999	−2.990	−2.900
f(x)	13.7100	13.0701	13.0070	12.9930	12.9301	12.3100

$\lim\limits_{x \to -3} f(x) = 13$

22.

x	3.900	3.990	3.999	4.001	4.010	4.100
f(x)	0.2516	0.2502	0.25002	0.24998	0.2498	0.2485

$\lim\limits_{x \to 4} f(x) = 0.25$

24.

x	10	100	1000
f(x)	−0.1230	−0.124 98	−0.124 999 8

$\lim\limits_{x \to \infty} f(x) = -0.125$

26. −5   28. $\frac{1}{9}$ 30. 2   32. −4   34. $\frac{3}{7}$ 36. 4   38. does not exist   40. 6   42. $\frac{1}{7}$

44. 0   46. 5.8, 5.98, 5.998, 6.2, 6.02, 6.002; $\lim\limits_{x \to 3} f(x) = 6$

48. 0.820 895 5, 0.530 223 5, 0.503 002 2; $\lim\limits_{x \to \infty} f(x) = 0.5$ 50. 92 cm   52. 5 Ω   54. 1

56. $\lim\limits_{x \to 0^+} 2^{1/x} = \infty$, $\lim\limits_{x \to 0^-} 2^{1/x} = 0$

Exercises 22-2, p. 655

2. (slopes) −1.75, −1.95,   4. (slopes) 1.75, 2,71,   6. −2   8. 3
   −1.995, −1.9995   2.97, 2.997

m = −2

m = 3

10. $m_{tan} = -x_1$; −2, 2   12. $m_{tan} = -6x_1$; 0, −12   14. $m_{tan} = 4x_1 - 4$; 0, 2

16. $m_{tan} = 2 - 6x_1$; 2, −1   18. $m_{tan} = 3 - 3x^2$; −9, 3, −9   20. $m_{tan} = -4x_1^3$; 0, −4, −32

22. $\frac{\Delta y}{\Delta x} = -4.2$, $m_{tan} = -4$ 24. $\frac{\Delta y}{\Delta x} = 21.91$, $m_{tan} = 21$

Exercises 22-3, p. 660

2. 6   4. −5   6. −2x   8. −12x   10. 2x + 4   12. 3 − x   14. 2 − 12x²

16. $-\dfrac{1}{(x + 1)^2}$ 18. $-\dfrac{1}{(x - 1)^2}$ 20. $-\dfrac{2x}{(x^2 + 1)^2}$ 22. x² + x + 1

24. $-\dfrac{1}{x^2} - \dfrac{2}{x^3}$ 26. 9 − 3x²; −3   28. 2x + $\dfrac{2}{x^2}$; $-\dfrac{7}{2}$ 30. $\dfrac{1}{2\sqrt{x - 2}}$ 32. $\dfrac{x}{\sqrt{x^2 + 3}}$

Exercises 22-4, p. 664

2. $m = 4$ 4. $m = 0$

6. $-3.00, -3.00, -3.00, -3.00, -3.00$
$\lim\limits_{t \to 4} v = -3$ m/s

8. $35.59, 35.345, 35.149, 35.1049, 35.100\ 49$
$\lim\limits_{t \to 0.5} v = 35.1$ m/s

10. $-3; -3$ m/s   12. $40 - 9.8t; 35.1$ m/s

14. $\dfrac{4}{(t + 2)^2}$ 16. $v_0 - at$

18. $\dfrac{1}{\sqrt{2t + 1}}$ 20. $-a$ 22. 100

24. 1510 m   26. $-32\ 000$ bacteria/h

28. $-\dfrac{48}{(t + 3)^2}; -\$1300$/year   30. 9.77×10^{-3} s/cell   32. -0.50 N/m

Exercises 22-5, p. 669

2. $12x^{11}$ 4. $-42x^5$ 6. $15x^4$ 8. $3x^2 - 4x$ 10. $12x - 6$ 12. $16x^3 - 2$
14. $52x^3 - 18x^2 - 1$ 16. $-2x^7 + 2x^2$ 18. 13 20. 49 22. 0 24. -30 26. $60 - 9.$
28. $v_0 + at$ 30. 0 32. 70 34. $0, 2$ 36. $12e$ 38. $3220d^2$
40. $kx(5x^3 + 1350x - 7000)$ 42. $\$132$/part   44. -13.3 cm^2

Exercises 22-6, p. 674

2. $12x^3 + 3$ 4. $42x^6 + 8x^3$ 6. $9x^2 + 2x + 3$ 8. $-24x^5 + 96x^3 + 6x^2 - 12$ 10. $8x^3 +$
12. $-72x^3 + 72x^2 + 18x - 20$ 14. $\dfrac{2}{(x + 1)^2}$ 16. $-\dfrac{1}{(2x + 3)^2}$ 18. $\dfrac{-12x + 10}{(3x^2 - 5x)^2}$
20. $\dfrac{-4x^3 + 24x^2}{(4 - x)^2}$ 22. $\dfrac{-48x^5 - 12}{(4x^5 - 3x - 4)^2}$ 24. $\dfrac{6x^4 - 30x^3 + 38x^2 - 4}{(2x^2 - 5x + 4)^2}$ 26. 2 28. $\dfrac{15}{32}$
30. $\dfrac{8x^3 - 16x^2 + 8x + 1}{(x - 1)^2}$ 32. -61 34. $+$, undef., $+$, undef., $+$; no   36. 20 sales/wee
38. 0.24 g/s   40. 0.925 V/s   42. $\dfrac{2ab - aV + pV^3}{RV^3}$ 44. $\dfrac{2kf(\omega^2 - \omega f + a^2)}{(\omega^2 - 2\omega f + f^2 + a^2)^2}$

Exercises 22-7, p. 680

2. $\dfrac{1}{4x^{3/4}}$ 4. $-\dfrac{8}{x^5}$ 6. $-\dfrac{1}{2x^{3/2}}$ 8. $\dfrac{2x - 6}{x^4}$ 10. $-8(1 - 2x)^3$ 12. $288x(8x^2 - 1)^5$
14. $-9(1 - 6x)^{1/2}$ 16. $\dfrac{6}{(1 - 3x)^{3/2}}$ 18. $\dfrac{70x^6}{(3x^7 - 4)^{1/3}}$ 20. $\dfrac{8x^5}{(4x^6 + 2)^{2/3}}$
22. $(2x - 21x^2)(1 - 3x)^4$ 24. $\dfrac{-2x^2 + 3x - 2}{2(1 - 2x)^2(x - 1)^{1/2}}$ 26. $-\dfrac{2}{9}$ 28. 9 30. $-\dfrac{8}{(4x + 3)^2}$
32. $x = \dfrac{1}{2}$ 34. $-\dfrac{4}{3}$ 36. $-\$4500$/unit   38. -9.09×10^5/m
40. $\dfrac{x}{v_1(a^2 + x^2)^{1/2}} + \dfrac{x - c}{v_2[b^2 + (c - x)^2]^{1/2}}$ 42. $\dfrac{-\omega^2 LV}{[R^2 + (\omega L)^2]^{3/2}}$ 44. $\dfrac{x - 3}{\sqrt{x^2 - 6x + 34}}$

Exercises 22-8, p. 684

2. 2 4. $\dfrac{5x^4 + 1}{5}$ 6. $-\dfrac{x}{2y}$ 8. $\dfrac{9x^2 - 1}{4y^3}$ 10. $\dfrac{4x^3}{1 - 6y^2}$ 12. $\dfrac{y}{8 - x}$
14. $\dfrac{y + (3 + y^2)(x + 1)^2}{(x + 1)(1 - 2xy - 2x^2y)}$ 16. $\dfrac{2 + 2x - 3x^2y^2}{1 + 2x^3y}$ 18. $\dfrac{4x^3y}{6y(y^2 + 2)^2 - x^4}$ 20. $\dfrac{2(3y - 1)}{2y - 6x - 3}$
22. -4 24. $-\dfrac{27}{44}$ 26. 0.363 m/s   28. $-\dfrac{r}{r + h}$ 30. $-\dfrac{b^3 + 3bh^2}{h^3 + 3b^2h}$ 32. $0.638, 0.143$

46

Exercises 22-9, p. 688

2. $f'(x) = 3 - 4x^3$, $f''(x) = -12x^2$, $f'''(x) = -24x$, $f^{(4)}(x) = -24$, $f^{(n)}(x) = 0$ $(n \geq 5)$
4. $y' = 10x^4 + 10x$, $y'' = 40x^3 + 10$, $y''' = 120x^2$, $y^{(4)} = 240x$, $y^{(5)} = 240$, $y^{(n)} = 0$ $(n \geq 6)$
6. $f'(x) = 9(3x + 2)^2$, $f''(x) = 54(3x + 2)$, $f'''(x) = 162$, $f^{(n)}(x) = 0$ $(n \geq 4)$
8. $y' = (x - 1)^2(4x - 1)$, $y'' = 12x^2 - 18x + 6$, $y''' = 24x - 18$, $y^{(4)} = 24$, $y^{(n)} = 0$ $(n \geq 5)$
10. $-40x^3$ 12. $2 - \dfrac{3}{4x^{5/2}}$ 14. $\dfrac{-8}{(6x + 5)^{5/3}}$ 16. $\dfrac{60}{(3 - 4x)^{5/2}}$ 18. $480(4x + 1)^4$
20. $144x(11x^3 + 3)(2x^3 + 3)^2$ 22. $\dfrac{4}{(1 + x)^3}$ 24. $\dfrac{3x}{(1 - x^2)^{5/2}}$ 26. $\dfrac{8}{(x + 2y)^3}$
28. $\dfrac{2y(x - y)}{(x - 2y)^3}$ 30. 24 32. 576 34. 0 m/s, -9.6 m/s^2 36. 24.8 W/(m$^2 \cdot$h^2)

Review Exercises for Chapter 22, p. 689

2. 8 4. 0 6. $\dfrac{10}{3}$ 8. -6 10. 7 12. 0 14. 6 16. $4x - 3x^2$ 18. $\dfrac{4}{(1 - 4x)^2}$
20. $-\dfrac{1}{2x^{3/2}}$ 22. $56x^6 - 1$ 24. $-\dfrac{6}{x^3} - \dfrac{2}{x^{3/4}}$ 26. $\dfrac{-2(x^2 - x - 1)}{(x^2 + 1)^2}$ 28. $24x(2x^2 - 3)^5$
30. $-\dfrac{63}{(3x - 1)^4}$ 32. $(x - 1)^2(x^2 - 2)(7x^2 - 4x - 6)$ 34. $\dfrac{1}{(x^2 + 1)^{3/2}}$ 36. $\dfrac{x(1 - y^2)}{y(x^2 - 1)}$
38. $-15\,972$ 40. $-\dfrac{1}{36}$ 42. $\dfrac{-16}{(1 - 8x)^{3/2}}$ 44. $480(3x + 1)(6x + 5)^2$ 46. \$16 000
48. $-\dfrac{8}{27}$ 50. 50 m 52. $\dfrac{a}{\sqrt{v_0^2 + 2as}}$ 54. $\dfrac{2t(t + 2)}{(1 + t)^2}$; 1.78 rad/s 56. 560 000 m^3/h
58. $\dfrac{aI(2b - I)}{(b - I)^2}$ 60. $\dfrac{-15}{(0.5t + 1)^2}$ 62. -0.0320 g/min^2 64. $4\pi r - \dfrac{200}{r^2}$ 66. 397 km/h
68. 1.05 km, 0.50 km

Exercises 23-1, p. 696

2. $4x - y - 18 = 0$ 4. $3x + 4y - 25 = 0$ 6. $x - 3y + 28 = 0$ 8. $x + 3y - 6 = 0$

10. $16x - y - 28 = 0$, $x + 16y - 66 = 0$
12. at $\left(2, \dfrac{3\sqrt{21}}{5}\right)$: $6x + 5\sqrt{21}y - 75 = 0$, $25\sqrt{21}x - 30y - 32\sqrt{21} = 0$
   at $\left(2, -\dfrac{3\sqrt{21}}{5}\right)$: $6x - 5\sqrt{21}y - 75 = 0$, $25\sqrt{21}x + 30y - 32\sqrt{21} = 0$
14. $y = x - 4$ 16. $128x - 32y + 97 = 0$ 18. $\dfrac{152}{3}$ 20. $3x + 4y - 15 = 0$
22. $y - 1.60 = -0.40(x - 0.32)$ 24. $(0, -10)$, $(-6, 8)$

Exercises 23-2, p. 700

2. 1.280 776 4 4. $-3.414\ 213\ 6$ 6. 0.798 360 3 8. 1.707 106 8 10. 2.302 775 6
12. $-1.561\ 552\ 8$ 14. 1.134 724 1 16. 0.593 513 5 18. 3.267 645 3
20. (0.7080, 0.6450) 22. 2.2143 μF, 3.2143 μF, 4.2143 μF 24. 0.629 cm

Exercises 23-3, p. 705

2. 0.54, 68.2° 4. 7.01, -87.3° 6. 0.26, 128.7° 8. 2.00, 268.9° 10. 4.8 m/s, 34°

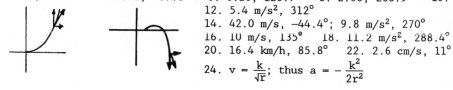

12. 5.4 m/s^2, 312°
14. 42.0 m/s, -44.4°; 9.8 m/s^2, 270°
16. 10 m/s, 135° 18. 11.2 m/s^2, 288.4°
20. 16.4 km/h, 85.8° 22. 2.6 cm/s, 11°
24. $v = \dfrac{k}{\sqrt{r}}$; thus $a = -\dfrac{k^2}{2r^2}$

Exercises 23-4, p. 709

2. 3000 J/s = 3 kJ/s   4. 0.098 Ω/min   6. −5200 km/h   8. −84.0 kPa/s
10. 3.25 cm²/s   12. 283 cm³/h   14. −39 kHz/s   16. 5.2×10⁻⁵ m/s³   18. 5.20 m/s
20. 10.2 m/s   22. 15.4 m/s   24. −0.978 cm/s

Exercises 23-5, p. 716

2. Inc. $x < 0$, dec. $x > 0$ 4. Inc. $-\sqrt{3} < x < 0$, $x > \sqrt{3}$, dec. $x < -\sqrt{3}$, $0 < x < \sqrt{3}$
6. Max. (0, 4)   8. Min. $(-\sqrt{3}, -9)$, $(\sqrt{3}, -9)$; Max. (0, 0)
10. Conc. down all x   12. Conc. up $x < -1$, $x > 1$; conc. down $-1 < x < 1$
14.   16.   18. Min. (0, −1)   20. Min. (5, −24)
                                        Conc. up all x   Max. (1, 8)
                                                              Infl. (3, −8)

22. Min. (2, −4)   24. Min. $(-\sqrt[3]{2}, 2 - 6\sqrt[3]{2})$ 26. Min. (2, −48)   28. Min. (0, 10)
    Max. (−2, 28)   Conc. up all x, $x \neq 0$ Max. (0,0)   Max. (4,74)
    Infl. (0, 12)   Infl. $(\sqrt[3]{2}, -18\sqrt[3]{4})$ Infl. (2, 42)

30.   32.   34.   36.

Exercises 23-6, p. 720

2. Asym. x=0, y=0   4. Int. (0, 0)   6. Int. $(-\sqrt[3]{4}, 0)$ 8. Min. (1, 4)
   Dec. x<0, x>0   Asym. x=2, y=1   Min. (2,3)   Max. (−1, −4)
                        Dec. x<2, x>2   Asym. x=0   Asym. x=0

10. Int. (0, 0)   12. Min. (0, 2)   14. Min. $\left(\frac{1}{4}, 3\right)$ 16. Infl. (1, 0)

    Max. $\left(3, \frac{3}{2}\right)$ Asym. x=−2, x=2,   Asym. x=0   Asym. x=0, x=2, y=0

    Min. $\left(-3, -\frac{3}{2}\right)$ y=0   Dom. x>0
    Asym. y=0

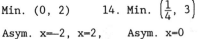

18. Min.(0, −1)   20. Max. (0, 100)   22. Max. (1, 9)   24. $\ell = x + \dfrac{40\,000}{x}$

Infl.$\left(\pm\dfrac{2}{\sqrt{3}},\ -\dfrac{1}{2}\right)$ Infl. (63,25)   Infl. (2, 8)   Min. (200, 400)

Asym. y−1   Int.(0,100),(89,0)   Asym. P = 0   Asym. x−0
Range −1≤y<1

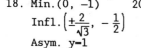

Exercises 23-7, p. 726

2. 200 barrels, $800   4. 50 mg   6. 2000 m   8. (1.2, 1.2)   10. 20, 20
12. 80 000 m^2 14. 30 m, 45 m   16. 1.41 cm, 1.41 cm   18. 108 dm^3 20. −15
22. 5.9 mm from 2.00 μC charge   24. h = 5.76 cm, r = 4.07 cm   26. 1250   28. 1.94 units
30. Let s − total distance, v − vel. of light, time = $\dfrac{s}{v}$

$\dfrac{s}{v} = \dfrac{\sqrt{a^2 + x^2} + \sqrt{(c - x)^2 + b^2}}{v}$; $\dfrac{d(s/v)}{dx} = 0$ leads to $\dfrac{x}{(a^2 + x^2)^{1/2}} = \dfrac{c - x}{[(c - x)^2 + b^2]^{1/2}}$

32. r − 3.17 m, h − 6.34 m

Review Exercises for Chapter 23, p. 728

2. 11x − 2y − 20 − 0   4. 32x − 2y − 191 − 0   6. x − 2y + 2 − 0   8. 1.26, 18.4°
10. −0.250 cm/s   12. 0.500 cm/s^2, 90°   14. 1.584 225 4   16. −0.585 786 4
18. Min. $\left(-\dfrac{1}{4},\ -\dfrac{9}{8}\right)$ 20. Min. $\left(-\dfrac{1}{3},\ \dfrac{23}{27}\right)$ 22. Min. (1, −11)   24.Min. (1, 4)

Conc. up all x   Max. (−1, 1)   Infl. (0, 0)   Max. (−1, −4)

Infl.$\left(-\dfrac{2}{3},\ \dfrac{25}{27}\right)$ (−2, 16)   Asym. x−0

26. 4x + y + 3 − 0   28. 1.20 m   30. 97.0 m/s   32. 0.31 Ω/min   34. 12 000 m
36.   38. 0.0185 A/min − 18.5 mA/min   40. r − 1, θ − 2   42. 9.45 m, 13.8 m
             44. Asym. R_T − 8   46. 0.14 cm^3/min   48. r − h − 0.56 m
             Dom. R ≥ 0
             Conc. down R ≥ 0

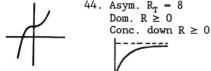

Exercises 24-1, p. 735

2. 6x dx   4. $\left(2 + \dfrac{1}{x^2}\right)dx$ 6. $\dfrac{dx}{(4 + 3x)^{2/3}}$ 8. $\dfrac{3x^2 dx}{2(1 - x^3)^{3/2}}$ 10. $\dfrac{(1 - 6x)dx}{\sqrt{1 - 4x}}$

12. $\dfrac{(3x - 4)dx}{(2x - 1)^{3/2}}$ 14. 2.595, 2.55   16. 10.885 571, 10.485 76   18. 3810, 3834

20. 0.002 655 3, 0.002 650 9   22. 0.8%   24. 0.104 m/s, 0.89%   26. 0.75%

28. $\dfrac{dF}{F} = -\dfrac{\dfrac{2k\,dr}{r^3}}{\dfrac{k}{r^2}} = -\dfrac{2\,dr}{r}$ 30. $\dfrac{dV}{V} = \dfrac{4\pi r^2 dr}{\dfrac{4}{3}\pi r^3} = \dfrac{3\,dr}{r}$ 32. 3.003 33

Exercises 24-2, p. 738

2. x^5 4. x^{10} 6. $2x^6 + x^2$ 8. $\frac{1}{3}x^3 - 5x$ 10. $x^{4/3}$ 12. $\frac{9}{4}x^{4/3} - 4x$ 14. $\frac{7}{5x^5}$

16. $\frac{2}{x^4}$ 18. $\frac{3}{4}x^4 - \frac{5}{3}x^3 - 3x$ 20. $\sqrt{x} + 4x$ 22. $\frac{1}{2}x^2 + x + \frac{1}{2x^2}$ 24. $(x^2 + 1)^3$

26. $(2x^4 + 1)^5$ 28. $-\frac{1}{16}(1 - x^2)^8$ 30. $(1 - x)^{5/4}$ 32. $\frac{1}{6}(4x + 3)^{3/2}$

Exercises 24-3, p. 744

2. $x^5 + C$ 4. $\frac{1}{6}x^6 + C$ 6. $\frac{3}{4}x^{4/3} + C$ 8. $2x^{1/2} + C$ 10. $x - \frac{3}{2}x^2 + C$

12. $\frac{4}{3}x^3 - x^2 + 5x + C$ 14. $2x^2 + \frac{1}{x^2} + C$ 16. $\frac{2}{5}x^{5/2} - \frac{5}{3}x^3 + C$

18. $\frac{3}{4}x^{4/3} + \frac{5}{6}x^{6/5} + \frac{7}{6}x^{6/7} + C$ 20. $x - x^2 + \frac{1}{3}x^3 + C$ 22. $\frac{1}{7}(x^3 - 2)^7 + C$

24. $\frac{3}{4}(1 - 2x)^{4/3} + C$ 26. $-\frac{6}{7}(1 - x^3)^{7/3} + C$ 28. $-\frac{1}{4}(4 - 3x)^{4/3} + C$

30. $\frac{2}{3}(2x^3 + 1)^{1/2} + C$ 32. $\frac{1}{27}\left(x^3 - \frac{3}{2}x^2\right)^9 + C$ 34. $y = \frac{1}{2}x^2 + x + \frac{9}{2}$

36. $y = \frac{1}{10}(x^4 - 6)^5 - 9990$ 38. $\frac{1}{9}(6x - 3)^{3/2} - 4$ 40. $v = 2t^{3/2} - 2t - 6$

42. $f = 160 - \frac{160}{(4 + L)^{1/2}}$ 44. $y = x^4 + 5$

Exercises 24-4, p. 750

2. 5, 5.6   4. 11, 13.695   6. 0.17, 0.189 375   8. 4.146, 4.540   10. 6   12. 15

14. $\frac{5}{24}$ 16. $\frac{14}{3}$

Exercises 24-5, p. 753

2. 8   4. $\frac{397}{5}$ 6. 2.67   8. $\frac{26}{3}$ 10. 24   12 10.7   14. $\frac{72}{5}$ 16. $\frac{5}{8}$

18. $\frac{2}{15}(841\sqrt{29} - 4\sqrt{2})$ 20. 2   22. 0.001 81   24. $\frac{56}{3}$ 26. 0   28. $\frac{11}{756}$ 30. −47.99

32. −0.7034   34. $\frac{1}{4}kR^4$ 36. 817 kN

Exercises 24-6, p. 757

2. $\frac{35}{54}, \frac{2}{3}$ 4. 12.66, $\frac{38}{3} = 12.67$ 6. 0.337   8. 3.28   10. 0.219   12. 13.36

14. 31.70   16. 200.054 m

Exercises 24-7, p. 760

2.(a) 11.1, (b) 12   4.(a) 3.3922, (b) 3.3934   6. 0.3365   8. 3.2396   10. 0.2187

12. 0.2154   14. 31.60   16. 1.1380 A

Review Exercises for Chapter 24, p. 761

2. $5x + x^3 + C$ 4. $\frac{1}{3}x^3 - \frac{1}{2}x^6 + C$ 6. 2   8. $\frac{7}{10}$ 10. $2x^{3/2} + x^{1/2} - \frac{1}{4}x + C$

12. 0.525   14. $-\frac{1}{24}(1 - 2x^2)^6 + C$ 16. $\frac{1}{4}(3x + 1)^{4/3} + C$ 18. $\frac{8}{3}$

20. $-\frac{9}{80}(1 - 5x^4)^{4/3} + C$ 22. $-\frac{2}{3}(6 + 9x - x^3)^{1/2} + C$ 24. $\frac{43\ 904}{3}$ 26. $\frac{-4\ dx}{(2x - 1)^3}$

28. $\frac{(x^2 + 6x + 4)dx}{(4 - x^2)^2}$ 30. 0.24   32. $y = \frac{1}{6}(x^2 + 1)^3 - \frac{10}{3}$

34.(a) $\frac{3}{2}x^2 + 2x + C_1$, (b) $\frac{1}{6}(3x + 2)^2 + C_2$, $C_1 = C_2 + \frac{2}{3}$ 36. $\frac{48}{5}$ 38. 46.2   40. 6.58

42. 1.25   44. 12.36   46. 12 W   48. Let $y = x^{1/n}$, then $\frac{dy}{y} = \frac{1}{n}\left(\frac{dx}{x}\right)$

50. $Q = \frac{k}{12R}(4Rr^3 - 3r^4 - R^4) + Q_0$ 52. 66.4 cm

Exercises 25-1, p. 769

2. -7.0 m/s   4. 0.84 mm   6. $v = \dfrac{30 + 40t}{1 + 2t}$ 8. 6.9 m   10. 37 m

12. $s = \frac{1}{3}[20(1 + 0.2t)^{5/2} - 10t - 14]$ 14. 0.015 C

16. $t = 16$ μs; direction of current changed between $t = 0$ and $t = 16$ μs   18. 9.9 nV

20. 0.0283 s = 28.3 ms   22. $\theta = \frac{1}{240}[(8t + 1)^{5/2} - 20t - 1]$ 24. 23°C

26. $y = \frac{k}{6}(x^6 + 2025x^4 - 14\,000x^3)$ 28. $r = 53.8\sqrt{\lambda}$

Exercises 25-2, p. 775

2. 4   4. 27   6. 1   8. $\frac{9}{4}\sqrt[3]{3}$ 10. $\frac{4}{3}$ 12. $\frac{41}{3}$ 14. $\frac{8}{3}$ 16. $\frac{1}{6}$ 18. 1   20. $\frac{19}{6}$

22. $\frac{512}{15}$ 24. $\frac{9}{2}$ 26. 8   28. $\frac{343}{6}$ 30. 0.074 C   32. $5000   34. 77.3 m²   36. 88.0 m³

Exercises 25-3, p. 782

2. $\frac{8}{3}\pi$ 4. 8π 6. $\frac{16}{15}\pi$ 8. $\frac{2}{15}\pi$ 10. $\frac{496}{15}\pi$ 12. $\frac{8}{9}\pi(4\sqrt[4]{2} - 1)$ 14. $\frac{32}{5}\pi$

16. $\frac{16}{3}\pi$ 18. $\frac{1024}{5}\pi$ 20. $\frac{32}{3}\pi$ 22. $\frac{8}{3}\pi$ 24. $\frac{96}{5}\pi$ 26. $\frac{6}{5}\pi$ 28. $V = \frac{4}{3}\pi r^3$

30. 23.0 m³   32. 0.824 m³

Exercises 25-4, p. 789

2. $\left(\frac{67}{15}, 0\right)$ 4. $\left(\frac{5}{6}, 0\right)$ 6. $\left(\frac{1}{38}, \frac{454}{38}\right)$ 8. $\left(\frac{29}{22}, \frac{18}{11}\right)$ 10. $\left(0, \frac{4a}{3\pi}\right)$ 12. $\left(\frac{8}{5}, \frac{16}{7}\right)$

14. $\left(\frac{6}{5}, \frac{3}{2}\right)$ 16. $\left(\frac{5}{3}, 0\right)$ 18. $\left(0, \frac{102}{43}\right)$ 20. $\left(\frac{3}{8}a, 0\right)$

22. On axis of disk, 0.29 cm from larger base   24. 3.25 m above lower base

Exercises 25-5, 794

2. 89, 2.84   4. 211, 3.10   6. $\frac{15}{2}$ 8. $\frac{1}{5}\sqrt{5}$ 10. $\frac{mb^2}{3}$ 12. $\frac{4\sqrt{15}}{9}$ 14. $\frac{1}{6}\sqrt{10}$

16. $\frac{2}{3}\sqrt{3}$ 18. $\frac{2}{5}\sqrt{10}$ 20. mr^2 22. 2 kg·m²   24. $\frac{1}{3}mL^2$

Exercises 25-6, p. 799

2. 24 N·cm   4. 1.7×10^{-16} J   6. 0.09k N·m   8. 3500 N·m   10. 5.88×10^5 N·m

12. 623 kJ   14. 39 200 N   16. 401 kN   18. 35 900 N   20. 3.4 kN

22. 1.16 kPa   24. $\frac{1}{3}\pi r^3$ 26. 15.3 km   28. 9.6 cm²

Review Exercises for Chapter 25, p. 802

2. 267 m   4. 50 m   6. 4.33 C   8. 12.8 μF   10. 40.8%   12. $\frac{27}{4}$ 14. $\frac{1}{15}$ 16. $\frac{8}{3}$

18. $\frac{729}{35}\pi$ 20. $\frac{8}{3}\pi$ 22. $\frac{4}{3}\pi(64 - 15\sqrt{15}) = 24.7$ cm³   24. $\left(\frac{4}{3}, -\frac{2}{3}\right)$ 26. $\left(0, \frac{7}{3}\right)$ 28. $\frac{2}{3}\sqrt{2}$

30. $\frac{1}{3}\sqrt{3}$ 32. 5.03×10^6 kg·N   34. 10 600 m³   36. 107 m²   38. 157 kN   40. $\frac{1}{12}mL^2$

Exercises 26-1, p. 809

2. $12\cos 4x$ 4. $-5\cos(3 - x)$ 6. $\sin(1 - x)$ 8. $-48x\sin(6x^2 + 5)$

10. $18\sin^2(2x + 1)\cos(2x + 1)$ 12. $-\dfrac{4\cos\sqrt{x}\,\sin\sqrt{x}}{\sqrt{x}}$ 14. $2x^2\cos 2x + 2x\sin 2x$

16. $5(\cos 2x^3 - 6x^3\sin 2x^3)$ 18. $\cos x\cos 4x - 4\sin x\sin 4x$

20. $4(x - \cos^2 x)^3(1 + 2\cos x \sin x)$ 22. $\dfrac{2\sin 4x - 8x\cos 4x - 12\cos 4x}{\sin^2 4x}$

24. $\dfrac{5(\cos x + x\sin x)}{\cos^2 x}$ 26. $4\cos^3 4x \sin 2x \cos 2x - 12\sin^2 2x \cos^2 4x \sin 4x$

28. $\dfrac{15\cos 5x - 5\cos^2 3x \cos 5x - 6\sin 5x \cos 3x \sin 3x}{(3 - \cos^2 3x)^2}$ 30. $x\cos x$ 32. $x^2 \sin x$

34. $\lim\limits_{\theta \to 0} \dfrac{\tan \theta}{\theta} - \lim\limits_{\theta \to 0} \dfrac{\sin \theta}{\theta} = 1$ 36. (a) $-0.841\ 471\ 0$, value of derivative
(b) $-0.841\ 498\ 0$, slope of secant line

38. Resulting curve is $y = -\sin x$ 40. $\dfrac{\cos 2y + \cos x \cos y}{2x\sin 2y + \sin x \sin y}$

42. $\dfrac{d^2 y}{dx^2} = -4\cos 2x = -4y$ 44. $2\sin x \cos x = \sin 2x$ 46. -0.059 48. -2.646

50. $-2400\pi \sin\left(120\pi t + \tfrac{1}{6}\pi\right)$ 52. $\dfrac{Ln^2\cos \theta}{d(n^2 - \sin^2\theta)^{3/2}}$

Exercises 26-2, p. 813

2. $9\sec^2(3x + 2)$ 4. $-18\csc^2 6x$ 6. $-\dfrac{\sec\sqrt{1 - x}\,\tan\sqrt{1 - x}}{2\sqrt{1 - x}}$

8. $2\csc(1 - 2x)\cot(1 - 2x)$ 10. $8x\tan(x^2)\sec^2(x^2)$ 12. $2\cot(1 - x)\csc^2(1 - x)$

14. $3\sec^3 x \tan x$ 16. $-8x\csc^2(2x^2)\cot(2x^2)$ 18. $3\sec 4x\,(1 + 4x\tan 4x)$

20. $\tfrac{1}{2}\sin 2x \sec x \tan x + \sec x \cos 2x = \sin x \tan x + \sec x \cos 2x$

22. $-\dfrac{4x\csc^2 4x + \cot 4x}{2x^2}$ 24. $\dfrac{2\tan 3x(6\sec^2 3x + 3\sin x^2 \sec^2 3x - x\tan 3x \cos x^2)}{(2 + \sin x^2)^2}$

26. $-2\csc 2x(\cot 2x - 2\csc 2x)$ 28. $x\sec^2 x + \tan x + 4\sec^2 2x \tan 2x$

30. $18\csc^2 3x \cot 3x(1 - \csc^2 3x)^2$ 32. $\dfrac{3\csc^2(x + y)}{2y\sin y^2 - 3\csc^2(x + y)}$ 34. $30\sec^3 2x \tan 2x\ dx$

36. $(-6x\csc^2 3x + 2\cot 3x)dx$ 38. (a) $2.882\ 474\ 7$, value of derivative
(b) $2.883\ 02$, slope of secant line

40. $-2\cot x \csc^2 x = -2\csc^2 x \cot x$ 42. 6.945

44. Substitution for $\dfrac{dy}{dx}$ and y gives

$\cos x(\cos x - 3\sin^2 x \cos x) + 3(\cos^3 x \tan x)\sin x - \cos^2 x = 0$
which can be reduced to $0 = 0$.

46. 2.5 A   48. -0.47 m

Exercises 26-3, p. 817

2. $\dfrac{-2}{\sqrt{2 - x^2}}$ 4. $\dfrac{-1}{\sqrt{2x - 4x^2}}$ 6. $\dfrac{-15}{\sqrt{1 - 25x^2}}$ 8. $\dfrac{-6x}{\sqrt{-(x^4 + 2x^2)}}$ 10. $-\dfrac{1}{x^2 - 2x + 2}$

12. $\dfrac{48x^3}{1 + 9x^8}$ 14. $2x\,\text{Arccos}\,x - \dfrac{x^2}{\sqrt{1 - x^2}}$ 16. $\dfrac{4(x^2 + 1)}{\sqrt{1 - 16x^2}} + 2x\,\text{Arcsin}\,4x$

18. $\dfrac{2x - (1 + 4x^2)\text{Arctan}\,2x}{x^2(1 + 4x^2)}$ 20. $\dfrac{2x\,\text{Arctan}\,x - 1}{\text{Arctan}^2 x}$ 22. $\dfrac{36\,\text{Arcsin}^3 3x}{\sqrt{1 - 9x^2}}$

24. $\dfrac{1}{2\sqrt{(2x - x^2)}\,\text{Arcsin}(x - 1)}$ 26. $\dfrac{2}{\sqrt{1 - 4x^2}\,\text{Arccos}^2 2x}$ 28. $\dfrac{1 + x}{\sqrt{1 - x^2}}$

30. $\dfrac{2x\sqrt{1 - (x + y)^2} - 1}{\sqrt{1 - (x + y)^2} + 1}$ 32. $\dfrac{-2}{\sqrt{(1 - 16x^2)}(1 - \text{Arcsin}\,4x)}$

34. (a) 0.8, value of derivative, (b) $0.799\ 968\ 0$, slope of secant line

36. $\left(\dfrac{-2x^4}{\sqrt{1 - x^4}} + 3x^2\,\text{Arccos}\,x^2\right)dx$ 38. Use same method as used to derive Eq.(26-12).

40. Let $y = \text{Arccsc}\,u$; solve for u; take derivatives; substitute.

42. $\dfrac{f}{r\sqrt{fr - f^2}}$ 44. $A = \text{Arccos}\dfrac{89 - x^2}{80}$, 0.20 rad/cm

Exercises 26-4, p. 821

2. $\sec^2 x \geq 1$ for all x 4. Max. $\left(\frac{1}{4}\pi,\ \sqrt{2}\right)$ 6. Max. $(0.94,\ 1.76)$, $(3.71,\ 0.37)$

8. $13x + 2y - 40.97 = 0$ Min. $\left(\frac{3}{4}\pi,\ -\sqrt{2}\right)$ Min. $(2.57,\ -0.37)$, $(5.35,\ -1.76)$

10. $1.165\ 561\ 2$ 12. -3

14. -4.2 cm/s

16. 0.657 W/min

18. 2.5 cm/s, $310°$

20. 17.4 cm/s^2, $176°$

22. 0.167 rad/s   24. 0.0345 rad/s   26. -0.024 W   28. $-\$832$ 30. $60°$ 32. 1.34 m

Exercises 26-5, p. 826

2. $\frac{1}{x}\log_2 e$ 4. $\frac{6x\ \log_7 e}{x^2 + 1}$ 6. $\frac{12x}{3x^2 - 1}$ 8. $2 \cot x$ 10. $\frac{2}{4x - 3}$ 12. $\frac{2(6x^2 - 1)}{2x^3 - x}$

14. $x + 2x \ln 2x$ 16. $\frac{1 - \ln x}{x^2}$ 18. $-2x \tan x^2$ 20. $\frac{3x + 2}{2x(x + 1)}$ 22. $\frac{1}{x(1 + \ln^2 2x)}$

24. $3 \ln^2 x + \ln^3 x$ 26. $\frac{1}{\sqrt{x^2 - 1}}$ 28. $\frac{x + 1}{2x(x + \ln 3x)^{1/2}}$ 30. $\frac{2x^2 y - 3y}{3x + xy \cos y}$

32. $\frac{3(2x^2 - 1)(x^2 - \ln 2x)^2}{x}$ 34. 1.9998, slope of secant line; 2 is value of derivative

36. $3 + 2 \ln x$ 38. $-0.008\ 32$ 40. $\frac{\sin 2x + 2x \cos 2x \ln 4x}{x}dx$ 42. 2.386

44. $(\sin x)^x(x \cot x + \ln \sin x)$ 46. $k(1 + \ln N)$ 48. $\frac{2ka}{x\sqrt{a^2 + x^2}}$

Exercises 26-6, p. 829

2. $-(\ln 3)3^{1-x}$ 4. $(2x \ln 10)10^{x^2}$ 6. $6xe^{x^2}$ 8. $8x^3 e^{2x^4}$ 10. $2xe^{2x}(x + 1)$

12. $2e^x\left(\cos\frac{1}{2}x + 2 \sin\frac{1}{2}x\right)$ 14. $\frac{e^x(x - 1)}{x^2}$ 16. $2\left(e^{x^2-1}\right)(x \cos 2x - \sin 2x)$

18. $\frac{2 + e^{2x} - 2xe^{2x}\ln 2x}{x(e^{2x} + 2)^2}$ 20. $3(3e^{2x} + x)^2(6e^{2x} + 1)$ 22. $\frac{-2e^{6/x}(x^2\sin x + 3 \cos x)}{x^2}$

24. $6x\left(2e^{x^2} + x^2\right)^2\left(2e^{x^2} + 1\right)$ 26. $\frac{8y \ln y\ e^{-2/x}}{x^2 y - 4x^2 e^{-2/x}}$ 28. $e^{x^2}(2x \ln \cos x - \tan x)$

30. $6e^{x+1}\sec^2 e^{x+1}$ 32. $\frac{3e^{3x}}{1 + e^{6x}}$ 34. (a) $7.389\ 056\ 1$, value of derivative

(b) $7.389\ 425$, slope of secant line

36. -0.06246 38. $-2e^{-x}\cos x + 2e^{-x}(\cos x - \sin x) + 2e^{-x}\sin x = 0$

40. $e^x + y'e^y = (1 + y')e^{x+y} = (1 + y')(e^x + e^y)$ 42. $-\alpha I_0 e^{-\alpha x}$ 44. $\frac{dn}{dt} = a(N - n)$

46. $\left[\frac{1}{2}(e^u + e^{-u})\right]^2 - \left[\frac{1}{2}(e^u - e^{-u})\right]^2 = \frac{1}{4}(e^{2u} + 2 + e^{-2u} - e^{2u} + 2 - e^{-2u})$

48. $\frac{d \sinh x}{dx} = \frac{d}{dx}\left[\frac{1}{2}(e^x - e^{-x})\right] = \frac{1}{2}(e^x + e^{-x})$; $\frac{d^2 \sinh x}{dx^2} = \frac{d}{dx}\left[\frac{1}{2}(e^x + e^{-x})\right] = \frac{1}{2}(e^x - e^{-x})$

(similar for cosh x) $= \sinh x$

Exercises 26-7, p. 833

2. Int. $(1,\ 0)$, Max. $\left(e, \frac{1}{e}\right)$ 4. Min. $(1,\ e)$ 6. Int. $(1,\ 0)$, Dec. $x > 0$

Infl. $\left(e^{3/2},\ \frac{3}{2e^{3/2}}\right)$ Asym $x=0$, $y=0$ Conc. up $x > 0$

Asym. $x=0$, $y=0$ Asym. $x=0$

8. Int.$(0, -1)$, Max.$(0, -1)$ 10. Int.$(0, 0)$, $(0, \pi)$, ...   12. Int.$(0, 1)$, MIn. $(0,$
   Conc. down all x   Max., min. $x - \frac{1}{4}\pi$. $\frac{5\pi}{4}$, ...   Conc. up all x

 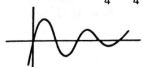

14. $2x - 5y + 3.54 = 0$ 16. $x + e^2y - 1 - e^4 = 0$ 18. $0.440\ 165\ 6$ 20. $\frac{dN}{dt} = -kN_0e^{-kt} = -k$

22. $R\left[\frac{E}{R}e^{-t/RC}\right] + \frac{1}{C}\left[CE(1 - e^{-t/RC})\right] = E$ 24. 1.5 dm   26. $\frac{di_m}{dt} = Ae^{k/T}(2T - k)\frac{dT}{dt}$

28. 169 h   30. -0.81 km/min   32. $\frac{1}{e}$

Review Exercises for Chapter 26, p. 835

2. $-12x^2\sec(1 - x^3)\tan(1 - x^3)$ 4. $-30\cos(1 - 6x)$ 6. $-10\cot 5x\ \csc^2 5x$

8. $\frac{3\sin^2\sqrt{x}\ \cos\sqrt{x}}{\sqrt{x}}$ 10. $2e^{\sin 2x}\cos 2x$ 12. $\frac{2x\cos x^2}{3 + \sin x^2}$ 14. $\frac{-4}{\sqrt{-(x^2 + 3x + 2)}}$

16. $\frac{\cos(\text{Arctan } x)}{1 + x^2}$ 18. $8\cos 2x(1 + \sin 2x)^3$ 20. $\cot 2x$ 22. $\frac{3[1 - \ln(3x + 1)]}{2(3x + 1)^2}$

24. $\frac{x - \sqrt{1 - x^2}\text{Arcsin } x}{4x^2\sqrt{1 - x^2}}$ 26. $-\frac{e^{\sqrt{1-x}}}{\sqrt{1 - x}}$ 28. $\frac{2\cos x}{3 + \sin x}$ 30. $e^{3x}\left(\frac{1}{x} + 3\ln x\right)$

32. $\frac{x + y}{x(2y - \ln 2x)}$ 34. $\frac{y(1 - 2x\ln y)}{x^2 - y}$ 36. $\frac{3(\ln 4x - \tan 4x)^2(1 - 4x\sec^2 4x)}{x}$

38. $\text{Arcsin}^2 x$ 40. $\frac{8x + 2}{1 + 4x^2}$ 42. Int.$(0, 1)$, Max.$\left(\frac{\pi}{2}, 3\right)$ 44. Int. $(0, 0)$, Asym. $x=-1$

Min.$\left(\frac{3}{2}\pi, -5\right)$ Conc. down $x > -1$

46. $1.73x + 3.00y - 0.48 = 0$
48. $2x + y - 2 - \frac{1}{4}\pi = 0$

 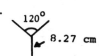

50. $\cos(x + 1) = \cos x\cos 1 - \sin x\sin 1$ 52. $1.676\ 018\ 6$ 54. 7.28 km/min
56. $\$67.65$/year   58. $\frac{kL}{x(x + L)}$ 60. $45°$ 62. $-\$4.40$ 64. 0.0028 ppm
66. $e^{-0.1t}(19.9\cos 100t - 100.02\sin 100t)$ 68. 370 m/s   70. $11.3°$
72. $\frac{di}{dt} = -\frac{R}{L}(i_0e^{-Rt/L}) = -\frac{R}{L}i$
74. 4.965 76.
78. Find derivatives and substitute.
80. 5570 cm^3

Exercises 27-1, p. 843

2. $\frac{1}{6}\cos^6 x + C$ 4. $\frac{3}{4}\sin^{4/3}x + C$ 6. $\frac{1}{4}\sec^4 x + C$ 8. $2(3^{3/4} - 1) = 2.559$

10. $-\frac{1}{10}(\text{Arccos } 2x)^5 + C$ 12. $\frac{1}{8}(\text{Arcsin } 4x)^2 + C$ 14. $\frac{1}{4}(3 + \ln 2x)^4 + C$ 16. 0

18. $\frac{4}{3}(1 - e^{-x})^{3/2} + C$ 20. $-\frac{1}{30}(1 + 3e^{-2x})^5 + C$ 22. $\frac{4}{5}(e^x + e^{-x})^{5/4} + C$

24. 0.449 26. $\frac{1}{8}\ln^2 9 = 0.6035$ 28. $y = \frac{1}{6}(1 + \tan 2x)^3 - 0.6745$ 30. πI 32. 36.6 km

Exercises 27-2, p. 846

2. $-\frac{1}{4}\ln|1 - 4x| + C$ 4. $-\frac{1}{3}\ln|1 - x^3| + C$ 6. $\frac{1}{2}\ln 41 = 1.857$ 8. $-\ln|\cos x| + C$

10. $\ln 1.25 = 0.223$ 12. $-\frac{2}{3}\ln|1 - e^{3x}| + C$ 14. $\ln|e^x + e^{-x}| + C$ 16. $\ln|1 - \cos^2 x| + C$

18. $2 \ln 6 = 3.58$ 20. $\frac{1}{2}\ln|1 + 2 \ln x| + C$ 22. $\frac{1}{2}\ln|x^2 + \sin 2x| + C$ 24. $-\dfrac{2}{1 + x^2} + C$

26. $-\frac{1}{x^2} - 3 \ln|x| + C$ 28. $\frac{1}{3} \ln \frac{7}{2} = 0.418$ 30. $\ln 2 = 0.6931$ 32. 10.2

34. $4 \ln 2 = 2.7726$ 36. $T = \dfrac{91.809}{r - 1}$ 38. $v = 20(1 - e^{-t})$ 40. $p = -\frac{3}{\pi}\ln(2 + \cos \pi t) + C$

Exercises 27-3, p. 850

2. $e^{x^4} + C$ 4. $-\frac{1}{2}e^{-4x} + C$ 6. $\frac{3}{4}(e^8 - e^4) = 2190$ 8. $-\frac{1}{2}e^{-x^2} + C$

10. $\frac{1}{2}(e^2 - 1) = 3.195$ 12. $e^{\tan x} + C$ 14. $\frac{1}{2}e^{2x} - 2x - \frac{1}{2}e^{-2x} + C$ 16. $\frac{3}{2}(e^2 - e) = 7.01$

18. $-\dfrac{4}{e^{\sin x}} + C$ 20. $\frac{1}{2}e^{\text{Arcsin } 2x} + C$ 22. $-\frac{1}{2}e^{2/x} + C$ 24. 0.904 26. $\int_a^b e^x dx = e^b - e^a$

28. $y = 2\left(\sqrt{e^{x+3}} - e^2\right)$ 30. $\frac{1}{3}(e - 1) = 0.573$ 32. $\dfrac{7}{\ln 2} = 10.10$ 34. $\frac{a}{T}(1 - e^{-I_0 T})$

36. $F = \frac{6}{\pi}(e^{\sin \pi t} - e^{-1})$

Exercises 27-4, p. 853

2. $4 \cos(2 - x) + C$ 4. $-\frac{1}{2}\csc 2x + C$ 6. $-\cot e^x + C$ 8. $2 \ln(1 - \cos 0.5) = -4.201$

10. $\frac{2}{3}\ln|\csc 3x - \cot 3x| + C$ 12. $\frac{3}{4}\ln|\csc 4x - \cot 4x| + C$ 14. $2(\sin e - \sin 1) = -0.8614$

16. $-2 \ln|\cos x| + C$ 18. $-\cot x + 2 \ln|\sin x| + C$ 20. $\tan x + \cot x + C$

22. $\ln|x + \tan x| + C$ 24. 1.865 26. 2 28. 2.644

30. $V = 584 \sin 377t$, $\sin 377t$ and $\cos 377t$ are $90°$ out of phase

32. $F = 2 \ln|\sec x + \tan x| + \sec x - 1$

Exercises 27-5, p. 858

2. $-\frac{1}{6}\cos^6 x + C$ 4. $3 \sin x - \sin^3 x + C$ 6. $\frac{1}{9}\cos^9 x - \frac{1}{7}\cos^7 x + C$ 8. $\frac{25}{168}\sqrt{2}$

10. $\frac{1}{2}x + \frac{1}{8}\sin 4x + C$ 12. $\frac{1}{8}(4 - \sin 4 \cos 4) = 0.4382$ 14. $-\frac{1}{2}\cot^2 x - \ln|\sin x| + C$

16. $-\frac{1}{16}\csc^4 4x + C$ 18. $-\frac{4}{3}\cot^3 x + 4 \cot x + 4x + C$ 20. $2\sqrt{\tan x}\left(\frac{1}{7}\tan^3 x + \frac{1}{3}\tan x\right) + C$

22. $\frac{1}{2}(\tan 2x - \cot 2x) + C$ 24. $\frac{1}{3}(\tan^3 x + \sec^3 x) + \tan x + C$ 26. 0.6188

28. $\frac{1}{6}\tan^6 x - \frac{1}{4}\tan^4 x + \frac{1}{2}\tan^2 x + \ln|\cos x| + C$ 30. 0.150 32. 1.317

34. $\int \sec^2 x \tan x \, dx = \frac{1}{2}\sec^2 x + C_1 = \frac{1}{2}\tan^2 x + C_2$; $C_2 = C_1 + \frac{1}{2}$ 36. $\tan \theta - \theta + C$

38. $\sqrt{2}$ 40. $kb\pi$

Exercises 27-6, p. 862

2. $\text{Arcsin} \frac{1}{7}x + C$ 4. $2 \text{Arctan} \frac{1}{2}x + C$ 6. $\text{Arcsin} \frac{2}{3} = 0.7297$

8. $\frac{1}{14}\left(\text{Arctan} \frac{6}{7} - \text{Arctan} \frac{2}{7}\right) = 0.0307$ 10. $-2(3 - 2x^2)^{1/2} + C$ 12. $\frac{3}{20}\text{Arctan} \frac{4}{5}x + C$

14. $\frac{1}{2}\text{Arctan} 1 = 0.3927$ 16. $\text{Arcsin}(\tan x) + C$ 18. $2 \text{Arctan}(x + 4) + C$

20. $\text{Arcsin}(x - 1) + C$ 22. 2.214 24. $\frac{3}{2}\text{Arctan} 2x - \frac{1}{4}\ln|1 + 4x^2| + C$

26. (a) logarithmic. (b) general power, (c) logarithmic
28. (a) inverse tangent, (b) general power, (c) logarithmic

30. $\frac{1}{6}\pi$ 32. 291 m^3 34. $s = t^2 - 8.49 \text{Arctan } 0.707t$ 36. $\frac{1}{2}\pi$

Exercises 27-7, p. 866

2. $-\frac{1}{2}x \cos 2x + \frac{1}{4}\sin 2x + C$ 4. $3e^x(x - 1) + C$ 6. $\frac{1}{4}\pi\sqrt{2} - \ln(1 + \sqrt{2}) = 0.229$

8. $x(\ln x - 1) + C$ 10. $\frac{2}{15}(x + 1)^{3/2}(3x - 2) + C$ 12. $\frac{1}{3}x^3 \ln 4x - \frac{1}{9}x^3 + C$

14. $\frac{1}{4}e^{2x}(2x^2 - 2x + 1) + C$ 16. $-\frac{1}{5}e^{-x}(\sin 2x + 2 \cos 2x) + C$ 18. $2\pi^2$ 20. 3.54

55

22. $y = \frac{1}{15}[2 + (3x^2 - 2)(1 + x^2)^{3/2}]$ 24. $\lim_{b \to \infty}[3.2 - 0.4(8 + b^2)e^{-b^2/8}] = 3.2$ nm

Exercises 27-8, p. 870

2. $\frac{x}{9\sqrt{x^2 + 9}} + C$ 4. $\sqrt{x^2 - 25} - 5$ Arcsec $\frac{1}{5}x + C$ 6. $\frac{3}{2}\ln\left|\frac{2 - \sqrt{4 - x^2}}{x}\right| + C$

8. $\frac{2}{3}(x^2 + 9)^{3/2} - 18\sqrt{x^2 + 9} + C$ 10. 0.0931   12. $\ln|x + 1 + \sqrt{x^2 + 2x}| + C$

14. 8 Arcsin$\frac{1}{4}x + \frac{1}{2}x\sqrt{16 - x^2} + C$ 16. $\frac{\tan x}{4\sqrt{4 - \tan^2 x}}$ 18. $\frac{\sqrt{8} - \sqrt{5}}{\sqrt{10}} = 0.187$

20. $\frac{2}{5}ma^2$ 22. 2.302   24. 37.8 cm^3

Exercises 27-9, p. 872

2. $4\left[\frac{1}{1 + x} + \ln(1 + x)\right] + C$ 4. $\frac{1}{4}\ln\frac{x - 2}{x + 2} + C$ 6. $\frac{5}{24}$ 8. x Arcsin 3x $+ \frac{1}{3}\sqrt{1 - 9x^2}$

10. $\frac{1}{3}(9x^2 + 16)^{3/2} + 16\sqrt{9x^2 + 16} - 64\ln\left[\frac{4 + \sqrt{9x^2 + 16}}{3x}\right] + C$ 12. $\tan x - x + C$

14. $\frac{5}{16}e^{4x}(4x - 1) + C$ 16. $-\frac{1}{8}\ln\frac{4x - 3}{4x + 3} + C$ 18. $\sqrt{4 + x^2} - 2\ln\left(\frac{2 + \sqrt{4 + x^2}}{x}\right) + C$

20. $\frac{1}{1 + 4x} - \ln\left(\frac{1 + 4x}{x}\right) + C$ 22. $\frac{1}{27}(26e^6 - 2)$ 24. $-\frac{2}{15}\cos^3 x(2 + 3\sin^2 x) + C$

26. $-\ln\left(\frac{1 - 4x}{x}\right) + C$ 28. undefined (div. by zero)   30. $\frac{1}{6}$Arcsec$\frac{1}{3}x^2 + C$

32. $-\frac{1}{30}(8 - 3x^4)(x^4 + 4)^{3/2} + C$ 34. $\frac{\pi}{8}(3e^4 + 1)$ 36. $\frac{3\sqrt{13}}{52} = 0.208$ 38. 0.0841 min

40. $F = \frac{kqQx}{b\sqrt{b^2 + x^2}} + C$

Review Exercises for Chapter 27, p. 874

2. $-\frac{1}{2}e^{\cos 2x} + C$ 4. $\frac{1}{2}(x^{4/3} + 1)^{3/2} + C$ 6. $\ln|2 + \tan x| + C$ 8. $\frac{1}{2}$Arcsin 2x $+ C$

10. $\frac{1}{6}(2\sqrt{2} - 1)$ 12. 1   14. $2\sqrt{\cos x}\left(\frac{1}{5}\cos^2 x - 1\right) + C$ 16. $\frac{3}{4}\ln|\sec 4x + \tan 4x| + C$

18. $x - \frac{1}{2}\sin 2x + C$ 20. $-8e^{-\sqrt{x}} - 2\sqrt{x} + C$ 22. $\frac{1}{4}\ln 3$ 24. $-\frac{1}{2}x\sqrt{9 - x^2} + \frac{9}{2}$Arcsin$\frac{1}{3}x +$

26. $\frac{1}{4}(4 + \ln 2x)^4 + C$ 28. $\frac{1}{8}(3x - 3\sin x \cos x - 2\sin^3 x \cos x) + C$

30. $\frac{1}{2}(x^2 + 1)$Arctan $x - \frac{1}{2}x + C$ 32. Arctan$(x + 3) + C$ 34. $\frac{1}{4}\pi$ 36. $\sin(\ln x) + C$

38. $\frac{1}{2}(1 + \ln x)^2 + C_1 = \ln x + \frac{1}{2}\ln^2 x + C_2$; $C_2 = C_1 + \frac{1}{2}$ 40. $y = -\frac{1}{3}(2 - e^x)^3 + \frac{13}{3}$

42. $y = \frac{1}{12}\left[12 + 2\sqrt{2} - \frac{(4 + x^2)^{3/2}}{x^3}\right]$ 44. $\ln 5 - \frac{4}{5} = 0.809$ 46. $\frac{506}{15}$ 48. $\frac{502}{15}\pi$

50. (1.65, 0.24)   52. 16π 54. $t = -[k_1\ln|a - x| + k_2\ln|b - x|]$ 56. $\frac{60}{\pi} = 19.1$ units

58. $\frac{A}{a}(e^a - e^{-a})$ 60. $\frac{1}{4}\pi^2 - 4$ 62. 0.441 dm   64. 1.10 m^2

Exercises 28-1, p. 880

2. $\frac{2}{3}, \frac{2}{9}, \frac{2}{27}, \frac{2}{81}$ 4. 1, $\frac{2}{3}$, 1, $\frac{10}{7}$ 6. (a) $\frac{3}{2}, \frac{5}{6}, \frac{7}{12}, \frac{9}{20}$, (b) $\frac{3}{2} + \frac{5}{6} + \frac{7}{12} + \frac{9}{20} + \ldots$

8. (a) $\frac{1}{6}, \frac{1}{12}, \frac{1}{20}, \frac{1}{30}$, (b) $\frac{1}{6} + \frac{1}{12} + \frac{1}{20} + \frac{1}{30} + \ldots$ 10. $a_n = \frac{1}{2^n}$ 12. $a_n = \left(-\frac{2}{3}\right)^n$

14. 1, 3, 8, 18, 35; divergent

16. 0.333 333 3, 0.222 222 2, 0.259 259 3, 0.246 913 6, 0.251 028 8; convergent; 0.25

18. 1, 1.333 333 3, 1.5, 1.6, 1.666 666 7; convergent; 1.7

20. 0.333 333 3, 0.733 333 3, 1.161 904 8, 1.606 349 2, 2.060 894 7; divergent

22. Convergent: $S = 2$ 24. Divergent   26. Convergent: $S = \frac{16}{3}$ 28. Convergent: $S = 64$

30. 656 nm, 486 nm, 434 nm, 365 nm   32. $r = -x$; $S = \frac{1}{1 + x}$

Exercises 28-2, p. 885

2. $x - \frac{1}{6}x^3 + \frac{1}{120}x^5 - \ldots$ 4. $x - \frac{1}{2}x^2 + \frac{1}{3}x^3 - \ldots$ 6. $1 + \frac{1}{3}x + \frac{2}{9}x^2 + \ldots$

8. $1 + \frac{1}{2}x^2 + \frac{1}{24}x^4 + \ldots$ 10. $x + x^2 + \frac{1}{3}x^3 + \ldots$ 12. $1 - 2x + 3x^2 - \ldots$

14. $1 + \frac{3}{2}x + \frac{3}{8}x^2 - \ldots$ 16. $4x - 8x^2 + \frac{64}{3}x^3 - \ldots$ 18. $1 - \frac{1}{2}x^4 + \ldots$

20. $1 + \frac{1}{2}x^2 + \ldots$ 22. $x + x^2 + \ldots$ 24. $1 - x^2 + \ldots$

26.(a) No, derivatives not defined at x=0, (b) Yes, f(x) and derivatives defined at x=0

28. $f(x) = (1 + x)^n = 1 + nx + \frac{n(n - 1)}{2}x^2 + \frac{n(n - 1)(n - 2)}{6}x^3 + \ldots$

30. $f^{(n)}(0) = 0$ except $f''(0) = 4$ and $f^{iv}(0) = 24$ 32. $c^2 = \frac{ab^2 + as^2 + bs^2}{a}$

Exercises 28-3, p. 891

2. $1 - 2x + 2x^2 - \frac{4}{3}x^3 + \ldots$ 4. $x^4 - \frac{1}{6}x^{12} + \frac{1}{120}x^{20} - \frac{1}{5040}x^{28} + \ldots$

6. $1 - \frac{1}{2}x^4 - \frac{1}{8}x^8 - \frac{1}{16}x^{12} - \ldots$ 8. $-x - \frac{1}{2}x^2 - \frac{1}{3}x^3 - \frac{1}{4}x^4 - \ldots$ 10. 0.3886

12. −0.007 48   14. $x + \frac{x^3}{3!} + \frac{x^5}{5!} + \frac{x^7}{7!} + \ldots$ 16. $x + \frac{1}{3}x^3 + \frac{2}{15}x^5 + \ldots$

18. $\frac{d}{dx}\left(1 + x + \frac{1}{2}x^2 + \frac{1}{6}x^3 + \ldots\right) = 1 + x + \frac{1}{2}x^2 + \ldots$

20. $-\int(1 + x + x^2 + \ldots)dx = -x - \frac{1}{2}x^2 - \frac{1}{3}x^3 - \ldots$ 22. 1   24. 1.53   26. 0.0294

28. $q = c(6at - 6a^2t^2 - 33a^3t^3 + 35a^4t^4)$

Exercises 28-4, p. 894

2. 0.625   4. 0.998 75   6. 0.6068   8. 0.069 756 5   10. −0.051 293   12. 0.918 24

14. 0.052 025   16. 1.303 34   18. 0.8903   20. 1.0442   22. 0.021   24. 0.002

26. 3.142   28.(a) 2.007 s, (b) 2.011 s

30. $f(p) = -p - \frac{1}{20}p^2 - \frac{1}{400}p^3 - \ldots$, −2.22 cm, −2.22 cm   32. 1.54%

Exercises 28-5, p. 898

2. 2.013   4. 1.87   6. 0.4695   8. 1.9628   10. $\frac{1}{2}\sqrt{2}\left[1 - \left(x - \frac{1}{4}\pi\right) - \frac{1}{2}\left(x - \frac{1}{4}\pi\right)^2 + \ldots\right]$

12. $\ln 3 + \frac{1}{3}(x - 3) - \frac{1}{18}(x - 3)^2 + \ldots$ 14. $\frac{1}{2} - \frac{1}{4}(x - 2) + \frac{1}{8}(x - 2)^2 + \ldots$

16. $-\frac{1}{2}\left(x - \frac{1}{2}\pi\right)^2 - \frac{1}{12}\left(x - \frac{1}{2}\pi\right)^4 - \frac{1}{45}\left(x - \frac{1}{2}\pi\right)^6 - \ldots$ 18. 1.131   20. 3.873   22. 1.036

24. 0.7431   26. 2.4265, 2.460 06, 2.459 603 1   28. 0.9189, 0.9651, 0.9657

Exercises 28-6, p. 905

2. $f(x) = \frac{4}{\pi}\left(\sin x + \frac{1}{3}\sin 3x + \frac{1}{5}\sin 5x + \ldots\right)$

4. $f(x) = \frac{1}{4} + \frac{1}{\pi}\left(\cos x - \frac{1}{3}\cos 3x + \ldots\right)$
$+ \frac{1}{\pi}\left(\sin x + \sin 2x + \ldots\right)$

6. $f(x) = 2\left(\sin x - \frac{1}{2}\sin 2x + \frac{1}{3}\sin 3x - \ldots\right)$

8. $f(x) = \frac{1}{3}\pi^2 - 4\left(\cos x - \frac{1}{4}\cos 2x + \frac{1}{9}\cos 3x - \ldots\right)$

10. $f(x) = \frac{1}{6}\pi^2 + 2\left(-\cos x + \frac{1}{4}\cos 2x - \ldots\right)$
$+ \frac{\pi^2 - 4}{\pi}\sin x - \frac{1}{2}\pi \sin 2x + \ldots$

12. $f(x) = 2 - \frac{16}{\pi^2}\left(\cos \frac{\pi x}{4} + \frac{1}{9}\cos \frac{3\pi x}{4} + \ldots\right)$

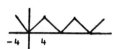

14. $L = f(t) = 15\pi + \frac{120}{\pi}\left(-\cos 2t - \frac{1}{9}\sin 6t + \ldots\right)$
$+ \frac{240}{\pi}\left(\sin t - \frac{1}{9}\sin 3t + \ldots\right)$

16. Ex. 9: $f(x) = f(-x)$, no sine terms; Ex. 6: $f(x) = -f(-x)$, no cosine terms

Review Exercises for Chapter 28, p. 907

2. $x + \frac{1}{6}x^3 + \frac{1}{120}x^5 + \ldots$ 4. $1 + 2x + 3x^2 + \ldots$ 6. $1 + \frac{1}{2}x^2 + \frac{1}{4}x^4 + \ldots$
8. $1 + x + x^2 + \ldots$ 10. 0.0953   12. 0.061 048 5   14. 1.504   16. 0.990 548 1
18. 4.02   20. 0.6820   22. 0.100 110 9   24. $\frac{1}{2}\ln\frac{1}{2} - \left(x - \frac{1}{4}\pi\right) - \left(x - \frac{1}{4}\pi\right)^2 - \ldots$
26. $f(x) = -\frac{1}{4}\pi - \frac{2}{\pi}\left(\cos x + \frac{1}{9}\cos 3x + \ldots\right)$
$- \left(\sin x - \frac{1}{2}\sin 2x + \ldots\right)$

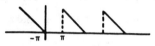

28. $f(x) = \frac{8}{\pi}\left(\sin \frac{\pi x}{3} + \frac{1}{3}\sin \pi x + \ldots\right)$

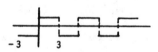

30. $2x - \frac{2}{3}x^3 + \frac{1}{20}x^5 - \ldots$ 32. 0.25, 0.5357, 0.8357, 1.1434, 1.4559, divergent
34. $f^{(n)}(0) = 0$, except $f(0) = c$, $f'(0) = b$, $f''(0) = 2a$
36. (a) $\sin 1 - \cos 1 = 0.301\ 17$, (b) 0.301 19
38. $\left(x^2 - \frac{1}{3}x^4 + \ldots\right) + \left(1 - x^2 + \frac{1}{4}x^4 + \frac{1}{12}x^4 + \ldots\right) = 1$
40. $1 - \frac{x^2}{2!} + \frac{x^4}{4!} + \ldots = 1 - \frac{(-x)^2}{2!} + \frac{(-x)^4}{4!} + \ldots$ 42. 0.003 05k
44. $\frac{m}{M} = k^2\omega^2 + \frac{1}{3}k^4\omega^4$ 46. $x = R - \sqrt{R^2 - y^2}$; $x = \frac{y^2}{2R} + \frac{y^4}{8R^3} + \frac{y^6}{16R^5} + \ldots$ 48. 2.0 km

Exercises 29-1, p. 912

(Note: "Answers" in this section are the unsimplified expressions obtained by substituting functions and derivatives in the given equation.)

2. general solution   4. particular solution   6. $2x = 2x$ 8. $(2x + 3) - 3 = 2x$
10. $\frac{2x}{(x^2 + c)^2} = 2x\left(-\frac{1}{x^2 + c}\right)^2$ 12. $(3x^2 + 1) - 3x^2 = 1$ 14. $-\frac{1}{2}\cos x + \frac{9}{2}\cos x = 4\cos x$
16. $c_3 e^x = c_3 e^x$ 18. $x(3cx^2 - 2x) - 3(cx^3 - x^2) = x^2$
20. $e^{-2x}(-8c_2 + 12 - 8x) + 4[e^{-2x}(4c_2 - 4 + 4x)] + 4[e^{-2x}(-2c_2 + 1 - 2x)] = 0$
22. $(x + x\ln x - cx) - x(1 + \ln x - c) = 0$ 24. $x\left(\frac{2c_2}{x^3}\right) + 2\left(-\frac{c_2}{x^2}\right) = 0$
26. $2xy\left(\frac{c - 2x}{2y}\right) + x^2 = cx - x^2$ 28. $x^4\left(-\frac{c}{x^2}\right)^2 - x\left(-\frac{c}{x^2}\right) = c^2 + \frac{c}{x}$

Exercises 29-2, p. 917

2. $4y^3 + 3x^4 = c$ 4. $xy = c$ 6. $\sqrt{1 + y^2} + \ln x = c$ 8. $\dfrac{2}{y^2} + \sqrt{1 + 4x^2} = c$

10. $y = e^{-x} + c$ 12. $y = -2x^2 + 3x + c$ 14. $y^2 = \text{Arctan}^2 x + c$

16. $\cos x + \sin y = c$ 18. $y = c - e^{\sin x}$ 20. $2x^3 + 3y^2 = c$

22. $(y^2 - 1)(x^3 + 1) = c$ 24. $(y + 1)(\sin x + 1) = c$ 26. $y = 3 - 2e^{-2x}$

28. $x = \sin y$ 30. $2 \ln y = x$ 32. $x = \ln\dfrac{\pi \sin^2 y}{2y}$

Exercises 29-3, p. 919

2. $y^2 + xy = c$ 4. $x - \dfrac{x}{y} = c$ 6. $x + \sin(xy) = c$ 8. $3 \ln xy + 4y^3 = c$

10. $2y + \ln(x^2 + y^2) = c$ 12. $y(x^2 + y^3) = 1 + c(x^2 + y^3)$ 14. $e^{x+y} + 2x^2 = c$

16. $yx^3 + x^2 = c$ 18. $\ln(x^2 + y^2) = 4x - 4$ 20. $y^3 + 3e^{x/y} = 11$

Exercises 29-4, p. 922

2. $y = e^{-3x}(x + c)$ 4. $y = e^{-x}(\sin x + c)$ 6. $y = \dfrac{1}{6}(5 + ce^{-3x})$ 8. $x^3(3y - 1) = c$

10. $2y = 9x + cx^{1/3}$ 12. $y = ce^{x^3/3} - 3$ 14. $y = \dfrac{1}{2}x(\ln^2 x + c)$

16. $y = \dfrac{1}{5}(2 \sin x - \cos x) + ce^{-2x}$ 18. $y = e^{2x}(2x + c)$ 20. $y \sec x = \ln \cos x + c$

22. $y = \dfrac{2}{3}x + cx^{-1/2}$ 24. $\sqrt{1 + x^2}(y - 1) = c$ 26. $y = \dfrac{1}{2}(5e^{4x} - 1)$ 28. $y = e^{4x} + e^{3x}$

30. $y(\csc x - \cot x) = \ln\dfrac{(\sqrt{2} - 1)(\csc 2x - \cot 2x)}{\csc x - \cot x}$ 32. $3y = f(x) + 10[f(x)]^{-2}$

Exercises 29-5, p. 927

2. $2xy + y^2 = 3$ 4. $y = e^{-x} + e^{-2x}$ 6. $3y^2 + x^2 = c$ 8. $y = cx$ 10. 1620 years

12. $N = \dfrac{r}{k}(1 - e^{-kt})$ 14. $v = \sqrt{\dfrac{2GM}{r_0 r}(r_0 - r)}$ 16. $P = xe^{-x^2}$ 18. $T = 20 + 80(0.375)^{t/10}$

20. 13.9 years   22. $i = 0.020e^{-15t}$ 24. $I = 2.0 - 2.0 \times 10^{-4}(100 - t)^2$ 26. 80 nC

28. 7.79 kg   30. After 2.3 s   32. $v = \dfrac{1 + 0.622e^{-1.4t}}{1 - 0.622e^{-1.4t}}$

34. $x = 2t + 4$, $y = \dfrac{10}{2t + 5}$ 36. 280 min   38. 7.03 min   40. $x^2 + y^2 = cy$

Exercises 29-6, 29-7, p. 933

2. $y = c_1 + c_2 e^{-x}$ 4. $y = c_1 e^{4x} + c_2 e^{-2x}$ 6. $y = c_1 e^{-x} + c_2 e^{-6x}$

8. $y = c_1 e^{x/2} + c_2 e^{-7x/2}$ 10. $y = c_1 e^{3x/4} + c_2 e^{-3x/2}$ 12. $y = c_1 e^{2x} + c_2 e^{3x/2}$

14. $y = e^{-x/2}\left(c_1 e^{\sqrt{21}x/2} + c_2 e^{-\sqrt{21}x/2}\right)$ 16. $y = e^{3x/4}\left(c_1 e^{\sqrt{17}x/4} + c_2 e^{-\sqrt{17}x/4}\right)$

18. $y = e^{x/10}\left(c_1 e^{\sqrt{61}x/10} + c_2 e^{-\sqrt{61}x/10}\right)$ 20. $y = e^{x/16}\left(c_1 e^{\sqrt{33}x/16} + c_2 e^{-\sqrt{33}x/16}\right)$

22. $y = -4 + 8e^{x/4}$ 24. $y = \dfrac{2}{e^{-5/2} - 1}(e^{-5x/2} - 1)$ 26. $y = c_1 + c_2 e^{x/2} + c_3 e^{-2x}$

28. $y = c_1 e^x + c_2 e^{-x} + c_3 e^{2x}$

Exercises 29-8, p. 937

2. $y = (c_1 + c_2 x)e^{3x}$ 4. $y = (c_1 + c_2 x)e^{-x/4}$ 6. $y = c_1 \sin x + c_2 \cos x$

8. $y = e^x(c_1 \sin\sqrt{3}x + c_2 \cos\sqrt{3}x)$ 10. $y = (a_1 + a_2 x)e^{3x/2}$ 12. $y = c_1 \sin\dfrac{2}{3}x + c_2 \cos\dfrac{2}{3}x$

14. $y = (c_1 + c_2 x)e^{4x/3}$ 16. $y = e^{2x}(c_1 \sin x + c_2 \cos x)$ 18. $y = e^{-2x}(c_1 \cos\sqrt{2}x + c_2 \sin\sqrt{2}x)$

20. $y = (c_1 + c_2 x)e^{-x/30}$ 22. $y = e^{5x/2}\left(c_1 e^{\sqrt{41}x/2} + c_2 e^{-\sqrt{41}x/2}\right)$ 24. $y = c_1 e^{5x/6} + c_2 e^{-5x/6}$

26. $y = 2\sqrt{3} \sin \frac{4}{3}x + 2 \cos \frac{4}{3}x$ 28. $y = -\frac{5}{2}xe^{-5x/2}$ 30. $(D^2 - 6D + 9)y = 0$

32. $(D^2 - 4D + 5)y = 0$

Exercises 29-9, p. 941

2. $y = c_1e^{3x} + c_2e^{-2x} + \frac{1}{9} - \frac{2}{3}x$ 4. $y = c_1e^{-x} + c_2e^x - 4 - x^2$

6. $y = e^{-x/4}\left(c_1\sin\frac{\sqrt{7}}{4}x + c_2\cos\frac{\sqrt{7}}{4}x\right) + \frac{1}{11}e^{2x}$ 8. $y = c_1e^x + c_2e^{-2x} - 5 - 2x - \frac{5}{8}e^{2x} + \frac{1}{2}xe^{2x}$

10. $y = c_1\sin 2x + c_2\cos 2x + \frac{1}{3}\sin x + 1$

12. $y = (c_1 + c_2x)e^x + 10 + 6x + x^2 - \frac{2}{25}\sin 3x + \frac{3}{50}\cos 3x$

14. $y = c_1e^{-6x} + c_2e^{x/2} - \frac{4}{3}x - \frac{22}{9}$ 16. $y = c_1\cos 2x + c_2\sin 2x - \frac{2}{5}\sin 3x$

18. $y = c_1e^{x/2} + c_2e^{-3x} + \frac{1}{4}e^x + \frac{4}{15}e^{2x}$

20. $y = e^{x/2}\left(c_1\sin\frac{\sqrt{3}}{2}x + c_2\cos\frac{\sqrt{3}}{2}x\right) + x + 1 + \cos x$ 22. $y = c_1e^{2x/3} + c_2e^{-x} - \frac{5}{2} - x + \frac{1}{2}e^x$

24. $y = e^{-4x}(c_1\sin\sqrt{17}x + c_2\cos\sqrt{17}x) - 130 - 16x - x^2 - \frac{4}{13}e^{-2x}$

26. $y = -\frac{225}{196}e^{x/3} + \frac{91}{700}e^{3x} + \frac{1}{35}xe^{-2x} + \frac{22}{1225}e^{-2x}$ 28. $y = (1 + 8x)e^x + (x - 3)e^{2x}$

Exercises 29-10, p. 947

2. $\theta = e^{-0.1t}(0.0032 \sin 3.13t + 0.10 \cos 3.13t)$ 4. 0.056 in.

6. $x = e^{-1.23t}(0.10 \cos 13.9t - 0.0068 \sin 13.9t)$

8. $x = e^{-1.23t}(0.019 \sin 13.9t + 0.10 \cos 13.9t) + 0.051 \sin 2t + 0.0013 \cos 2t$

10. $x = e^{-1.83t}(0.0458 \sin 5.99t + 0.150 \cos 5.99t)$ 12. $q = 10^5\cos 10^5t$

14. $i = 3.16 \sin 316t$ 16. $\frac{20}{229}(2 \sin 200t - 15 \cos 200t)$

18. $i = e^{-75t}(5.00\times10^{-3}\cos 3530t + 0.46\times10^{-3}\sin 3530t) - 1.60\times10^{-5}e^{-100t}$

20. $i = 4.02\times10^{-2}\cos 10t + 1.61\times10^{-4}\sin 10t$

Exercises 29-11, p. 953

2. $L(f) = \int_0^\infty e^{-st}e^{-at}dt = \int_0^\infty e^{-(s+a)t}dt$ 4. $L(f) = \int_0^\infty e^{-st}te^{-at}dt = \int_0^\infty te^{-(s+a)t}dt$

6. $\frac{4}{s(s^2 + 4)}$ 8. $\frac{8}{s^2 + 6s + 25}$ 10. $\frac{s^5 + 3s^4 + 30s^3 + 126s^2 + 201s + 243}{(s^2 + 9)^2(s^2 + 6s + 10)}$

12. $\frac{6}{s^4} - \frac{3}{(s + 1)^2}$ 14. $(s^2 - 3s)L(f) - 2s + 7$ 16. $(s^2 - 3s + 2)L(f) + s - 5$

18. $\frac{3}{2}\sin 2t$ 20. $\frac{3}{8}(2t - \sin 2t)$ 22. $t \cos t$ 24. $\frac{1}{3}e^{-2t}(3 \cos 3t + \sin 3t)$

Exercises 29-12, p. 956

2. $y = 2e^{2t}$ 4. $y = \frac{1}{2}(1 - e^{-2t})$ 6. $y = \frac{1}{2}t^2e^{-2t}$ 8. $y = e^{2t} + e^{-2t}$ 10. $y = -2te^{-t}$

12. $y = \frac{1}{2}e^{-t/2}(t + 2)$ 14. $\frac{1}{4}(2t - \sin 2t)$ 16. $\frac{1}{8}(\sin 2t - 2t \cos 2t)$

18. $\theta = 0.089 \sin 4.5t$ 20. $i = 0.1(1 - e^{-40t})$ 22. $q = 2500t^2e^{-1000t}$

24. $y = \frac{1}{2}e^{-3t}(\cos t + 3 \sin t)$

Review Exercises for Chapter 29, p. 957

2. $e^y = e^x + c$ 4. $2xy = y^2 + c$ 6. $y = c_1 e^{2x} + c_2 e^{x/2}$ 8. $y = e^{-x}(c_1 \sin x + c_2 \cos x)$

10. $\ln^2 x = 2 \ln y + c$ 12. $y = e^x(1 - x) + ce^{2x}$ 14. $\sin y - y \cos y = \ln x - \frac{1}{x} + c$

16. $y = e^{x/2}(c_1 + c_2 x)$ 18. $2x^2 = 4y - \ln^2 y + c$ 20. $\ln(x^2 + y^2) = 2x + c$

22. $y = (c_1 + c_2 x)e^{-3x} + \frac{x}{3} - \frac{2}{9}$ 24. $y = c_1 \sin \frac{3}{2}x + c_2 \cos \frac{3}{2}x + \frac{1}{13}xe^x - \frac{8}{169}e^x$

26. $y = c_1 \sin x + c_2 \cos x - \frac{4}{3}\cos 2x$ 28. $y = c_1 + c_2 e^{-x} + \frac{1}{2}e^x + \frac{1}{10}\sin 2x - \frac{1}{5}\cos 2x$

30. $3y^3 - 29y + 6x = 0$ 32. $3y = 2ye^{-x} + 2xye^{-x} + 2$ 34. $y = -\frac{44}{13}e^{-2x} + \frac{70}{13}e^{3x/5}$

36. $y = -e^x(1 + 2x) + 2 + x + e^{2x}$ 38. $y = -4 + 5e^{t/2}$ 40. $y = e^{-2t}(t + 2)$

42. $y = e^{-2t}(\cos t + 3 \sin t)$ 44. $y = e^{-2t}(t^2 + t)$ 46. $56.3°C$ 48. $v = 380(1 - e^{-0.026t})$

50. $p = cV^k$ 52. $m = m_0(1 - e^{-kt})$ 54. $x^2 + y^2 = c$ 56. $i = 0.5(1 - e^{-20t})$

58. $y = 3e^{-0.1t}\cos 63.3t$ 60. $q = 10^{-6}(\sin 4470t - 2 \cos 4470t + 2e^{-200t})$

62. $q = \dfrac{CE_0}{1 - \omega^2 LC}\left(\sin \omega t - \omega\sqrt{LC}\, \sin\dfrac{t}{\sqrt{LC}}\right)$ 64. $i = 15e^{-5t}$

66. $q = 0.01e^{-6t}(\cos 100t + 0.06 \sin 100t)$ 68. $y = \cos 3.16t$

70. $\theta = \theta_0 \cos \omega t + \dfrac{\omega_0}{\omega}\sin \omega t$; $\omega = \sqrt{\dfrac{2k}{mr^2}}$ 72. 2.47 L

Exercises S-1, p. 646 (BTM), p. 964 (BTMC)

2. $x = -1, y = 3$ 4. $x = -\frac{14}{11}, y = \frac{1}{11}$ 6. $x = 2, y = \frac{1}{2}, z = -1$

8. $x = \frac{1}{2}, y = -\frac{1}{6}, z = \frac{2}{3}$ 10. $x = 1, y = -1, z = 2, t = -3$

12. Unlimited: $x = 1, y = 0, z = 2$; $x = \frac{1}{2}, y = -5, z = 4$

14. Unlimited: $x = 2, y = -1, z = 1$; $x = 3, y = -2, z = 2$ 16. Inconsistent
18. $x = -2, y = 2, z = 3$

20. Unlimited: $x = \frac{5}{4}, y = -\frac{9}{2}, z = -1, t = 5$; $x = \frac{11}{8}, y = -\frac{3}{4}, z = 0, t = -\frac{1}{2}$

22. $0.5\,\Omega, 3.0\Omega$ 24. $\$4600, \$4000, \$3400$

Exercises S-2, p. 651 (BTM), p. 969 (BTMC)

2. circle
   $x'^2 + y'^2 = 16$

4. hyperbola
   $11x'^2 - 14y'^2 = 8$

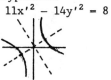

6. ellipse
   $x'^2 + 4y'^2 = 16$

8. parabola
   $y'^2 = 16x'$

10. hyperbola
    $2x'^2 - 3y'^2 = 6$

12. ellipse
    $x'^2 + 4y'^2 - 4x' - 8y' + 4 = 0$
    $x''^2 + 4y''^2 - 4 = 0$

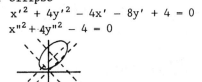

Exercises S-3, p. 972

2. $A = \frac{2V}{r} + 2\pi r^2$ 4. $T = F + 100h$ 6. -2 8. $6xt + xt^2 + t^3$

10. $2hx - 2kx - 2hy + h^2 - 2hk - 4h$ 12. $81z^6 - 9z^5 - 2z^3$ 14. $y \le 1$ 16. 150 Pa

18. 28.7 m³   20. $f = \dfrac{0.16}{\sqrt{LC}}$

Exercises S-4, p. 978

2. 4. 6. 8. 10.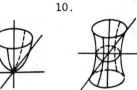

12. 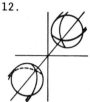 14. 16. 18. 20.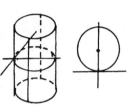

Exercises S-5, p. 983

2. $\frac{\partial z}{\partial x} = \frac{2x}{y} - 2y$, $\frac{\partial z}{\partial y} = -\frac{x^2}{y^2} - 2x$ 4. $\frac{\partial f}{\partial x} = -\frac{\sin x}{1 - \sec 3y}$, $\frac{\partial f}{\partial y} = -\frac{3(2 + \cos x)\sec 3y \tan 3y}{(1 - \sec 3y)^2}$

6. $\frac{\partial z}{\partial x} = 4(2x + y^3)(x^2 + xy^3)^3$, $\frac{\partial z}{\partial y} = 12xy^2(x^2 + xy^3)^3$ 8. $\frac{\partial y}{\partial r} = \frac{2r}{r^2 + s}$, $\frac{\partial y}{\partial s} = \frac{1}{r^2 + s}$

10. $\frac{\partial z}{\partial x} = \frac{3y + x^2y - 2x\sqrt{1 - x^2y^2}\,\text{Arcsin}\,xy}{(3 + x^2)^2\sqrt{1 - x^2y^2}}$, $\frac{\partial z}{\partial y} = \frac{x}{(3 + x^2)\sqrt{1 - x^2y^2}}$

12. $\frac{\partial f}{\partial x} = e^x(\cos xy - y \sin xy) - 2e^{-2x}\tan y$, $\frac{\partial f}{\partial y} = -xe^x\sin xy + e^{-2x}\sec^2 y$ 14. $2e$

16. $\frac{\partial^2 f}{\partial x^2} = \frac{2(3x^2 - 1)(2 + \cos y)}{(1 + x^2)^3}$, $\frac{\partial^2 f}{\partial y^2} = \frac{-\cos y}{1 + x^2}$, $\frac{\partial^2 f}{\partial x \partial y} = \frac{\partial^2 f}{\partial y \partial x} = \frac{2x \sin y}{(1 + x^2)^2}$ 18. 0.807

20. $\left(\frac{R_2}{R_1 + R_2}\right)^2$ 22. $f_s\left[\frac{f_s(v + v_0)}{(v - v_s)^2}\right] = \left[f_s\left(\frac{v + v_0}{v - v_s}\right)\right]\left(\frac{f_s}{v - v_s}\right)$

24. $-5e^{-t}\sin 4x = \frac{1}{16}(-80e^{-t}\sin 4x)$

Exercises S-6, p. 988

2. $-\frac{14}{5}$ 4. $\frac{844}{15}$ 6. $\frac{1}{4}(2 - \sqrt{3}) = 0.0670$ 8. -0.658 10. 8π 12. 18

14. 50.3 m^3 16.

Exercises S-7, p. 992

2. $\ln\left|\frac{x^2}{x + 1}\right| + C$ 4. $\frac{17}{2}\ln|2x - 1| - 8\ln|x - 1| + C$ 6. $\frac{1}{2}x^2 - 3x + \ln\left|\frac{(x + 2)^8}{x + 1}\right| + C$

8. 0.0958 10. $\frac{1}{4}\ln\left|\frac{x^4(2x + 1)^3}{2x - 1}\right| + C$ 12. $\frac{1}{60}\ln\left|\frac{(x + 2)^3(x - 3)^2}{(x - 2)^3(x + 3)^2}\right| + C$ 14. 0.1633 N·m

16. $2\pi \ln\frac{6}{5} = 1.146$

Exercises S-8, p. 997

2. $\frac{1}{x} + \ln\left|\frac{x-1}{x}\right| + C$ 4. 3.172   6. $\ln\left|\frac{x+1}{x-1}\right| - \frac{2x}{x^2-1} + C$ 8. $\ln|x - 2| + \text{Arctan } x + C$

10. $\frac{1}{4}\ln(4x^2 + 1) + \ln|x^2 + 6x + 10| + \text{Arctan}(x + 3) + C$

12. $\ln|x + 1| - \frac{1}{2}\ln|x^2 + 1| + \text{Arctan } x + \frac{x}{1 + x^2} + C$ 14. $\pi \ln\frac{25}{9} = 3.210$ 16. $190 \ \mu\text{C}$

Exercises B-1, p. A-9

2. kW, $1 \text{ kW} = 10^3 \text{ W}$ 4. ps, $1 \text{ ps} = 10^{-12}$ s   6. gigaohm, $1 \text{ G}\Omega = 10^9 \ \Omega$
8. picofarad, $1 \text{ pF} = 10^{-12}$ F   10. 10^6 mg   12. 0.800 kPa   14. 304.65 K
16. 1.75×10^4 cm²   18. 3.60×10^{-10} cm³   20. 125 mL   22. 0.200 m/s   24. 2.20×10^3 cm/min
26. 5.10×10^{-6} kg/mm³   28. 1.50×10^{-2} mC/V   30. 101.3 kPa   32. 1.4×10^7 mg
34. 0.0042 kL/s   36. 1.27×10^8 m/h²   38. 5.52×10^3 kg/m³   40. 0.135 J/(s·cm²)

Exercises B-2, p. A-12

2. 298 000 is approximate   4. 10 and 15 are approximate; 5.50 is exact
6. 21 is exact; 81.6 is approximate   8. First 50 is exact; second 50 is approximate
10. 2; 2   12. 4; 2   14. 4; 5   16. 3; 1   18.(a) same, (b) 7.673
20.(a) 70 370, (b) 70 370   22.(a) 37.1, (b) same   24.(a) same, (b) 50.060
26.(a) 80.5, (b) 81   28.(a) 31 500, (b) 31 000   30.(a) 9560, (b) 9600
32.(a) 0.735, (b) 0.74   34.(a) 81.9, (b) 82   36.(a) 10.0, (b) 10
38. 81.5 L; 82.5 L   40. 6.25 mm

Exercises B-3, p. A-16

2. 179   4. 186.6   6. 0.35   8. 2.04   10. 1.8   12. 18.48   14. 0.000600
16. 177.4   18. 2.5   20. 0.263   22. 8.1   24. 8.2   26. 87.39   28. 18.759
30. 14.882   32. 0.004 99   34. 27.0 N   36. 0.159 W   38. 849 Mg
40. 35 m/s   42. 7.08 km
44. First reading is not precise to thousandths. Therefore, there cannot be a
    difference to this precision. First reading should probably be 0.020 A.

Exercises for Appendix C, p. A-22

2. 125°   4. 130°   6. 128°   8. 80°   10. 110°   12. 124°   14. 90°   16. 65°
18. 160°   20. 80°   22. 15   24. 26   26. 3.464   28. 8.938   30. 15.50   32. 36.13
34. 9   36. 16   38. 11.7 cm, 7.62 cm, 50°, 100°, 30°   40. $4\sqrt{7} = 10.6$
42. 22 cm   44. 21 hm   46. 51.2 mm   48. 62.48 m   50. 41.4 mm   52. 22.2 cm
54. 21 dm²   56. 48.72 cm²   58. 235 hm²   60. 26.7 cm²   62. 366 m³   64. 90 dm³
66. 572 cm³   68. 5.52 cm³   70. 1283 cm³   72. 2030 dm³   74. 407 mm²
76. 428.8 dm²   78. 735 cm²   80. 9830 mm²   82. 19.6 cm²   84. 4010 m³

Exercises for Appendix D, p. A-31

(Most answers have been rounded off to four significant digits.)

2. 1061.8   4. −7.1405   6. 24.71   8. 693.3   10. 0.9722   12. 7.769   14. 0.001 645
16. 41.92   18. 0.9128   20. −0.4863   22. 0.5954   24. −1.133   26. 2.552
28. 4.227°   30. −0.045 18   32. 4.256   34. 0.8100   36. 5.323   38. 1.595   40. 383.1
42. 6.313×10^{20} 44. 1.076   46. 0.8411   48. −1.251   50. 558.2   52. −2.088
54. 3.658   56. 3122   58. 1.044   60. 94.82°   62. -3.083×10^{-3} 64. 2.781   66. −61.16°
68. 0.8210   70. 123.5   72. 1.000   74. −0.5962   76. 5.031   78. 7.001   80. 574.2
82. 8.686   84. −0.052 82   86. 2.170   88. 0.515 04   90. 0.060 83   92. 0.999 998 9

2. 150 IF X < > 0 THEN LET Y = 8/X
   or 145 IF X = 0 THEN 190
      150 LET Y = 8/X

6. One way is to find the sum of
   two of the vectors, and then use
   the program again to find the
   sum of the resultant and the
   third vector. Another way is to
   change lines 130, 180, and 190,
   and in a line 175 include
   magnitude C and angle TC.

12. $N = 0$; $C_1=1$, $C_2=C_3=C_4=0$
    $N = 1$: $C_1=A$, $C_2=B$, $C_3=C_4=0$
    $N = 2$: $C_1=A^2$, $C_2=2AB$, $C_3=B^2$, $C_4=0$

16. 130 LET F=3*X^3 − X^2 − 4*X + 1
    140 LET D=9*X^2 − 2*X − 4

20. 100 DEF FN F(N) = 1/5^N
    150 FOR N=0 TO I

4. 210 IF C1*B2−C2*B1 < > 0 THEN PRINT
   "SYSTEM IS INCONSISTENT"
   215 IF C1*B2−C2*B1 = 0 THEN PRINT
   "SYSTEM IS DEPENDENT"

8. 125 INPUT "C = ";C
   a200 HPLOT X,80−26*A*SIN(B*X/19+C)
   m200 p=80−26*a*sin(b*x/19+c)
   m205 line(x,p)−(x,p)

10. 185 PRINT 1/A; " LOG ";B;" = LOG ";
    B^(1/A)

14. Change line 20, omit line 240
    180 LET Y1=8*A*X/B:LET Y2=−8*A*X/B
    250 LET Y3=8*A*SQR(X^2+B^2)/B
    260 LET Y4=−8*A*SQR(X^2+B^2)/B

18. For $y=x^n$ in the interval $0<x<1$,
    values of y decrease as n in-
    creases for a given value of x.
    Thus, volumes decrease as n
    increases.

ANSWERS TO ODD-NUMBERED EXERCISES

Exercises 1-1, page 5

1. Integer, rational, real; irrational, real **3.** Imaginary; irrational, real **5.** $3, \frac{7}{2}$ **7.** $\frac{6}{7}, \sqrt{3}$ **9.** $6 < 8$
11. $\pi > -1$ **13.** $-4 < -3$ **15.** $-\frac{1}{3} > -\frac{1}{2}$ **17.** $\frac{1}{3}, -\frac{1}{2}$ **19.** $-\frac{\pi}{5}, \frac{1}{x}$
21. **23.**

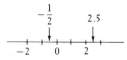 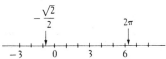

25. $-18, -|-3|, -1, \sqrt{5}, \pi, |-8|, 9$
27. (a) Positive integer, (b) negative integer, (c) positive rational number less than 1
29. (a) To right of origin, (b) to left of -4 **31.** L, t variables; a constant **33.** $N = 1000an$ **35.** Yes

Exercises 1-2, 1-3, page 11

1. 11 **3.** -11 **5.** 9 **7.** -3 **9.** -24 **11.** 35 **13.** -3 **15.** 20 **17.** 40 **19.** -1
21. 9 **23.** Undefined **25.** 20 **27.** -5 **29.** -9 **31.** 24 **33.** -6 **35.** 3
37. Commutative law of multiplication **39.** Distributive law **41.** Associative law of addition
43. Associative law of multiplication **45.** Positive **47.** $\frac{4}{2} = 2; \frac{2}{4} = \frac{1}{2}; 2 \neq \frac{1}{2}$
49. 100 m + 200 m = 200 m + 100 m; commutative law of addition **51.** 8($2000 + $1000); distributive law

Exercises 1-4, page 13

1. 3.62 **3.** -0.118 **5.** 0.1356 **7.** 6.086 **9.** $43.011 = 43.011$, commutative law of addition
11. $478.303\,41 = 478.303\,41$, distributive law **13.** -8.21 **15.** 0.976 **17.** 14.9 **19.** 0.0330
21. (a) 0.36, (b) 0.36 **23.** No, $-0.000\,007\,3$ **25.** 0.242 424 2, (b) 3.141 592 7 **27.** Error
29. 2.6 g/day **31.** 59.14%

Exercises 1-5, page 19

1. x^7 **3.** $2b^6$ **5.** m^2 **7.** $\dfrac{1}{n^4}$ **9.** a^8 **11.** t^{20} **13.** $8n^3$ **15.** a^2x^8 **17.** $\dfrac{8}{b^3}$ **19.** $\dfrac{x^8}{16}$ **21.** 1

23. 3 **25.** $\frac{1}{6}$ **27.** s^2 **29.** $-t^{14}$ **31.** $64x^{12}$ **33.** 1 **35.** b^2 **37.** $\frac{1}{8}$ **39.** 1 **41.** $\dfrac{a}{x^2}$

43. $\dfrac{x^3}{64a^3}$ **45.** $64g^2s^6$ **47.** $\dfrac{5a}{n}$ **49.** -53 **51.** -10 **53.** 6230 **55.** -0.421 **57.** $\dfrac{r}{6}$ **59.** 68.9 W

Exercises 1-6, page 23

1. 45 000 **3.** 0.002 01 **5.** 3.23 **7.** 18.6 **9.** 4×10^4 **11.** 8.7×10^{-3} **13.** 6×10^0
15. 6.3×10^{-2} **17.** 5.6×10^{13} **19.** 2.2×10^8 **21.** 3.2 kg **23.** 0.91 ms or 910 μs **25.** 5.3×10^6 m
27. 2.64×10^{-2} g **29.** 4.85×10^{10} **31.** 5.0350×10^1 **33.** 6.5×10^6 kW **35.** 0.000 000 000 001 6 W
37. 200 000 **39.** 3×10^{-6} W **41.** 0.000 000 000 000 000 000 000 000 000 000 000 000 000 24
43. 3.6×10^4 km **45.** 2.57×10^{14} cm^2 **47.** 3.433 Ω

Exercises 1-7, page 26

1. 5 **3.** -11 **5.** -7 **7.** 20 **9.** 5 **11.** -6 **13.** 5 **15.** 31 **17.** $3\sqrt{2}$ **19.** $2\sqrt{3}$
21. $2\sqrt{11}$ **23.** $30\sqrt{7}$ **25.** $4\sqrt{21}$ **27.** 7 **29.** 10 **31.** $3\sqrt{10}$ **33.** 9.24 **35.** 0.6877 **37.** 2.66 s
39. 35.4 m **41.** 48.3 cm **43.** No, not true if $a < 0$

Exercises 1-8, page 31

1. $8x$ **3.** $4x + y$ **5.** $a + c - 2$ **7.** $-a^2b - a^2b^2$ **9.** $4s + 4$ **11.** $-v + 5x - 4$ **13.** $5a - 5$
15. $-5a + 2$ **17.** $-2t + 5u$ **19.** $7r + 8s$ **21.** $19c - 50$ **23.** $3n - 9$ **25.** $-2t^2 + 18$ **27.** $6a$
29. $2a\sqrt{xy} + 1$ **31.** $4c - 6$ **33.** $8p - 5q$ **35.** $-4x^2 + 22$ **37.** $4a - 3$ **39.** $-6t + 13$
41. $2D + d$ **43.** $22 - 10x$

Exercises 1-9, page 33

1. a^3x **3.** $-a^2c^3x^3$ **5.** $-8a^3x^5$ **7.** $2a^8x^3$ **9.** $a^2x + a^2y$ **11.** $-3s^3 + 15st$ **13.** $5m^3n + 15m^2n$
15. $-3x^2 - 3xy + 6x$ **17.** $a^2b^2c^5 - ab^3c^5 - a^2b^3c^4$ **19.** $acx^4 + acx^3y^3$ **21.** $x^2 + 2x - 15$
23. $2x^2 + 9x - 5$ **25.** $6a^2 - 7ab + 2b^2$ **27.** $6s^2 + 11st - 35t^2$ **29.** $2x^3 + 5x^2 - 2x - 5$
31. $x^3 + 2x^2 - 8x$ **33.** $x^3 - 2x^2 - x + 2$ **35.** $x^5 - x^4 - 6x^3 + 4x^2 + 8x$ **37.** $2a^2 - 16a - 18$
39. $2x^3 + 6x^2 - 8x$ **41.** $4x^2 - 20x + 25$ **43.** $x^2 + 6ax + 9a^2$ **45.** $x^2y^2z^2 - 4xyz + 4$
47. $2x^2 + 32x + 128$ **49.** $-x^3 + 2x^2 + 5x - 6$ **51.** $6x^4 + 21x^3 + 12x^2 - 12x$ **53.** $x^2 + x - 2$ mm^2
55. $R^2 - r^2$

Exercises 1-10, page 37

1. $-4x^2y$ **3.** $4t^4/r^2$ **5.** $4x^2$ **7.** $-6a$ **9.** $a^2 + 4y$ **11.** $t - 2rt^2$ **13.** $q + 2p - 4q^3$
15. $\dfrac{2L}{R} - R$ **17.** $\dfrac{1}{3a} - \dfrac{2b}{3a} + 1$ **19.** $x^2 + a$ **21.** $2x + 1$ **23.** $x - 1$ **25.** $4x^2 - x - 1, R = -3$
27. $x + 5$ **29.** $x^2 + x - 6$ **31.** $2x^2 + 4x + 2, R = 4x + 4$ **33.** $x^2 - 2x + 4$ **35.** $x - y$
37. $A + \dfrac{\mu^2 E^2}{2A} - \dfrac{\mu^4 E^4}{8A^3}$ **39.** $5x + 16$

Exercises 1-11, page 40

1. 9   **3.** -1 **5.** 10   **7.** -5 **9.** -3 **11.** 1   **13.** $-\frac{7}{2}$ **15.** 8   **17.** $\frac{10}{3}$ **19.** $-\frac{13}{3}$ **21.** 2
23. 0   **25.** 9.5   **27.** -1.5 **29.** True for all values of x, identity   **31.** Not true for any x, contradiction
33. 1.3 km/h   **35.** 750 L

Exercises 1-12, page 43

1. $\dfrac{b}{a}$ **3.** $\dfrac{4m - 1}{4}$ **5.** $\dfrac{c + 6}{a}$ **7.** $2a + 8$ **9.** $\theta - kA$ **11.** $\dfrac{E}{I}$ **13.** $\dfrac{P}{2\pi f}$ **15.** $\dfrac{p - p_a}{dg}$ **17.** $\dfrac{PL^2}{\pi^2 I}$
19. $\dfrac{s + 16t^2}{t}$ **21.** $\dfrac{C_0^2 - C_1^2}{2C_1^2}$ **23.** $\dfrac{a + PV^2}{V}$ **25.** $\dfrac{Q_1 + PQ_1}{P}$ **27.** $\dfrac{N + N_2 - N_2 T}{T}$
29. $\dfrac{L - \pi r_2 - 2x_1 - x_2}{\pi}$ **31.** $\dfrac{gJP + V_1^2}{V_1}$ **33.** 56.3 cm   **35.** 204 K

Exercises 1-13, page 48

1. \$138, \$252   **3.** 22, 35, 50   **5.** 20 hectares at \$20 000/hectare, 50 hectares at \$10 000/hectare   **7.** 20 girders
9. $-2.3\ \mu A, -4.6\ \mu A, 6.9\ \mu A$ **11.** 90 cm, 135 cm   **13.** 6.9 km (main), 9.5 km (others)
15. 351.1 s, first car   **17.** 1600 km/h, 2000 km/h   **19.** 900 m   **21.** 6 L   **23.** 110 Mg, 140 Mg

Review Exercises for Chapter 1, page 50

1. -10 **3.** -20 **5.** 2   **7.** -25 **9.** -4 **11.** 5   **13.** $4r^2t^4$ **15.** $-\dfrac{6m^2}{nt^2}$ **17.** $\dfrac{z^6}{y^2}$ **19.** $\dfrac{8t^3}{s^2}$
21. $3\sqrt{5}$ **23.** $2\sqrt{5}$ **25.** 18.12   **27.** -1.440 **29.** $-a - 2ab$ **31.** $5xy + 3$ **33.** $2x^2 + 9x - 5$
35. $x^2 + 16x + 64$ **37.** $hk - 3h^2k^4$ **39.** $7a - 6b$ **41.** $13xy - 10z$ **43.** $2x^3 - x^2 - 7x - 3$
45. $-3x^2y + 24xy^2 - 48y^3$ **47.** $-9p^2 + 3pq + 18p^2q$ **49.** $\dfrac{6q}{p} - 2 + \dfrac{3q^4}{p^3}$ **51.** $2x - 5$ **53.** $x^2 - 2x + 3$
55. $4x^3 - 2x^2 + 6x, R = -1$ **57.** $15r - 3s - 3t$ **59.** $y^2 + 5y - 1, R = 4$ **61.** 4   **63.** $-\frac{9}{2}$ **65.** $-\frac{7}{3}$
67. 3   **69.** $-\frac{19}{5}$ **71.** 1   **73.** 4×10^4 km/h   **75.** 1.02×10^9 Hz   **77.** 1.2×10^{-6} cm^2
79. 0.000 18 kg/m^3 **81.** $\dfrac{5a - 2}{3}$ **83.** $\dfrac{4 - 2n}{3}$ **85.** $\dfrac{R}{n^2}$ **87.** $\dfrac{I - P}{Pr}$ **89.** $\dfrac{dV - m}{dV}$ **91.** $\dfrac{2C - mN_2}{m}$
93. $J - E + ES$ **95.** $\dfrac{d - 3kbx^2 + kx^3}{3kx^2}$ **97.** 1.81×10^{-2} m   **99.** 1.45 N   **101.** $1 - 4r + 3r^2$
103. $9x^2 - 1600x + 80\,000$ **105.** 16 384 bytes, 65 536 bytes   **107.** 1900 Ω, 3100 Ω
109. 52 vertical strands, 82 horizontal strands   **111.** After 1.4 h   **113.** 400 L   **115.** 27 m^2

Exercises 2-1, page 58

1. $A = \pi r^2$ **3.** $c = 2\pi r$ **5.** $A = 5\ell$ **7.** $A = s^2; s = \sqrt{A}$ **9.** $3, -1$ **11.** $11, 3.8$ **13.** $-18, 70$
15. $\frac{5}{2}, -\frac{1}{2}$ **17.** $\frac{1}{4}a + \frac{1}{2}a^2, 0$ **19.** $3s^2 + s + 6, 12s^2 - 2s + 6$ **21.** $62.9, 260$ **23.** $0.019\,88, -0.2998$
25. Square the value of the independent variable and add 2.   **27.** Cube the value of the independent variable
and subtract this value from 6 times the value of the independent variable.   **29.** $y = f(x), f(x) = x^2$
31. $P = f(p), f(p) = 40(p - 24)$ **33.** 3.4 m   **35.** 13.2 m, $0.4v + 0.032v^2$, 40.8 m, 40.8 m

Exercises 2-2, page 62

1. Domain: all real numbers; range: all real numbers   **3.** Domain: all real numbers except 0;
range: all real numbers except 0   **5.** Domain: all real numbers except 0; range: all real numbers $f(s) > 0$
7. Domain: all real numbers $h \geq 0$; range: all real numbers $H(h) \geq 1$ **9.** All real numbers $y > 2$
11. All real numbers except 2 and -4 **13.** $-10, \frac{8}{9}$ **15.** $-16, \frac{1}{2}$ **17.** $-1, \sqrt{5}$ **19.** $\frac{1}{2}, 0$
21. $d = 80 + 55t$ **23.** $w = 5500 - 2t$ **25.** $w = 0.5h - 390$ **27.** $C = 5\ell + 250$ **29.** $y = 3000 - 0.25x$

31. $A = \dfrac{1}{16}p^2 + \dfrac{(60 - p)^2}{4\pi}$ **33.** $C > 0$, with some upper limit depending on circuit

35. $w = \begin{cases} 110 & \text{for } h \leq 1000 \text{ m} \quad (w \text{ in kg}) \\ 0.5h - 390 & \text{for } h > 1000 \text{ m} \end{cases}$

Exercises 2-3, page 66

1. $(2, 1), (-1, 2), (-2, -3)$ **3.**

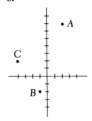

5. Isosceles triangle

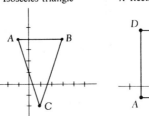

7. Rectangle

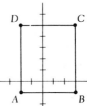

9. $(5, 4)$ **11.** $(3, -2)$ **13.** On a line parallel to the y-axis, one unit to the right
15. On a line parallel to the x-axis, three units above   **17.** On a line bisecting the first and third quadrants   **19.** 0
21. To the right of the y-axis   **23.** To the left of a line which is parallel to the y-axis and one unit to its left
25. On the negative y-axis   **27.** First, third

Exercises 2-4, page 71

1.

3.

5.

7.

9.

11.

13.

15.

17.

19.

21.

23.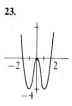

25.

27.

29.

31.

33.

35.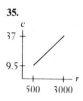

37.

39.

41. $A = 30w - w^2$

43.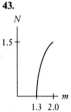

45.

47.

49. Yes

51. No

Exercises 2-5, page 75

1.

3.

5.

7.

9. (a) 132.1°C, (b) 0.7 min   **11.** (a) 7.8 V, (b) 43°C   **13.** 0.30 H   **15.** 7.2%   **17.**

19. 1.3 m³/s   **21.** 0.34   **23.** 76 m²   **25.** 130.3°C   **27.** 3.7 m³/s

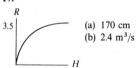

(a) 170 cm
(b) 2.4 m³/s

Exercises 2-6, page 79

1. 2   **3.** $-\frac{9}{2}$ **5.** 3.5   **7.** 0.5   **9.** 0.0, -1.0 **11.** -1.7 **13.** 0.7   **15.** 2.5   **17.** -2.8, 1.8
19. -1.6, 2.1   **21.** -2.0, 0.0, 2.0   **23.** 0.0, 1.3   **25.** 3.5   **27.** No real zeros   **29.** 6 V   **31.** 5.1 N
33. 18 cm, 30 cm   **35.** 67 m

Review Exercises for Chapter 2, page 80

1. $V = 8\pi r^2$ **3.** $y = -\frac{10}{9}x + \frac{250}{9}$ **5.** 16, -47 **7.** -5, -1.08 **9.** 3, $\sqrt{1 - 4h}$ **11.** $6hx + 3h^2 - 2h$
13. -3.67, 16.7   **15.** 0.165 72, $-0.215 66$ **17.** Domain: all real numbers; range: all real numbers $f(x) \geq 1$
19. Domain: all real numbers $t > -4$; range: all real numbers $g(t) > 0$

21.   **23.**   **25.**   **27.**   **29.**   **31.**

33. -0.5 **35.** $0, 4$ **37.** $-1.5, 1.0$ **39.** $-2.4, 0.0, 2.4$ **41.** $-1.2, 1.2$ **43.** 0 **45.** 0.4
47. $0.2, 5.8$ **49.** 1.4 **51.** $-0.7, 0.7$ **53.** Either a or b is positive, the other is negative **55.** 13.4

57.   **59.**   **61.** 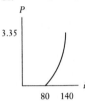 **63.**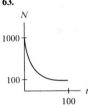

65. 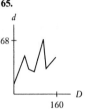 **67.** $f(10) = 204$ 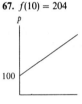 **69.** $33°C$ **71.** 6.5 h

Exercises 3-1, page 87

1.   **3.**   **5.** $405°, -315°$ **7.** $210°, -510°$ **9.** $430°30', -289°30'$
11. $638.1°, -81.9°$ **13.** $15.2°$ **15.** $82.91°$ **17.** $15.2°$
19. $86.05°$ **21.** $301.27°$ **23.** $-96.13°$ **25.** $47°30'$
27. $19°45'$ **29.** $-5°37'$ **31.** $24°55'$

33.   **35.**   **37.**   **39.** 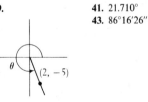 **41.** $21.710°$
43. $86°16'26''$

Exercises 3-2, page 91

1. $\sin\theta = \frac{4}{5}$, $\cos\theta = \frac{3}{5}$, $\tan\theta = \frac{4}{3}$, $\cot\theta = \frac{3}{4}$, $\sec\theta = \frac{5}{3}$, $\csc\theta = \frac{5}{4}$
3. $\sin\theta = \frac{8}{17}$, $\cos\theta = \frac{15}{17}$, $\tan\theta = \frac{8}{15}$, $\cot\theta = \frac{15}{8}$, $\sec\theta = \frac{17}{15}$, $\csc\theta = \frac{17}{8}$
5. $\sin\theta = \frac{40}{41}$, $\cos\theta = \frac{9}{41}$, $\tan\theta = \frac{40}{9}$, $\cot\theta = \frac{9}{40}$, $\sec\theta = \frac{41}{9}$, $\csc\theta = \frac{41}{40}$
7. $\sin\theta = \frac{\sqrt{15}}{4}$, $\cos\theta = \frac{1}{4}$, $\tan\theta = \sqrt{15}$, $\cot\theta = \frac{1}{\sqrt{15}}$, $\sec\theta = 4$, $\csc\theta = \frac{4}{\sqrt{15}}$
9. $\sin\theta = \frac{1}{\sqrt{2}}$, $\cos\theta = \frac{1}{\sqrt{2}}$, $\tan\theta = 1$, $\cot\theta = 1$, $\sec\theta = \sqrt{2}$, $\csc\theta = \sqrt{2}$
11. $\sin\theta = \frac{2}{\sqrt{29}}$, $\cos\theta = \frac{5}{\sqrt{29}}$, $\tan\theta = \frac{2}{5}$, $\cot\theta = \frac{5}{2}$, $\sec\theta = \frac{\sqrt{29}}{5}$, $\csc\theta = \frac{\sqrt{29}}{2}$
13. $\sin\theta = 0.846$, $\cos\theta = 0.534$, $\tan\theta = 1.58$, $\cot\theta = 0.631$, $\sec\theta = 1.87$, $\csc\theta = 1.18$
15. $\sin\theta = 0.1521$, $\cos\theta = 0.9884$, $\tan\theta = 0.1539$, $\cot\theta = 6.498$, $\sec\theta = 1.012$, $\csc\theta = 6.575$

17. $\frac{5}{13}, \frac{12}{5}$ **19.** $\frac{1}{\sqrt{2}}, \sqrt{2}$ **21.** 0.882, 1.33   **23.** 0.246, 3.94   **25.** $\sin \theta = \frac{4}{5}, \tan \theta = \frac{4}{3}$

27. $\tan \theta = \frac{1}{2}, \sec \theta = \frac{\sqrt{5}}{2}$ **29.** $\sec \theta$ **31.** $\dfrac{x}{y} \times \dfrac{y}{r} = \dfrac{x}{r} = \cos \theta$

Exercises 3-3, page 95

1. $\sin 40° = 0.64$, $\cos 40° = 0.77$, $\tan 40° = 0.84$, $\cot 40° = 1.19$, $\sec 40° = 1.30$, $\csc 40° = 1.56$
3. $\sin 15° = 0.26$, $\cos 15° = 0.97$, $\tan 15° = 0.27$, $\cot 15° = 3.73$, $\sec 15° = 1.04$, $\csc 15° = 3.86$
5. 0.381   **7.** 1.58   **9.** 0.9626   **11.** 0.99   **13.** 0.4085   **15.** 1.57   **17.** 1.32   **19.** 0.070 63
21. 70.97°   **23.** 65.70°   **25.** 11.7°   **27.** 49.453°   **29.** 53.44°   **31.** 81.79°   **33.** 74.1°   **35.** 17.85°
37. 0.8885   **39.** 0.936 14   **41.** 0.326   **43.** 2.356   **45.** 40°   **47.** 83°   **49.** 0.550   **51.** 0.880
53. 72.8°   **55.** 35°10′   **57.** 87 dB   **59.** 70.4°

Exercises 3-4, page 100

1.
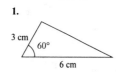

3.
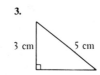

5. $B = 12.2°$, $b = 1450$, $c = 6850$
7. $A = 25.8°$, $B = 64.2°$, $b = 311$
9. $A = 57.9°$, $a = 20.2$, $b = 12.6$
11. $A = 21°$, $a = 32$, $B = 69°$

13. $a = 30.21$, $B = 57.90°$, $b = 48.16$ **15.** $A = 52.15°$, $B = 37.85°$, $c = 71.85$
17. $A = 52.5°$, $b = 0.661$, $c = 1.09$ **19.** $A = 15.82°$, $a = 0.5239$, $c = 1.922$
21. $A = 65.886°$, $B = 24.114°$, $c = 648.46$ **23.** $a = 3.3621$, $B = 77.025°$, $c = 14.974$ **25.** 4.45   **27.** 40.24°
29. $A = 52°20′$, $b = 0.684$, $c = 1.12$ **31.** $A = 58°40′$, $a = 143$, $B = 31°20′$
33. $a = c \sin A$, $b = c \cos A$, $B = 90° - A$ **35.** $A = 90° - B$, $b = a \tan B$, $c = a/\cos B$

Exercises 3-5, page 103

1. 97 m   **3.** 44.0 m   **5.** 0.4°   **7.** 850.1 cm   **9.** 765 cm   **11.** 26.6°, 63.4°, 90.0°   **13.** 57.4 m
15. 1.29 m   **17.** 8.1°   **19.** 1840 km   **21.** 651 m   **23.** 30.2°   **25.** 47.3 m   **27.** 200 m

Review Exercises for Chapter 3, page 105

1. 377.0°, $-343.0°$ **3.** 142.5°, $-577.5°$ **5.** 31.9°   **7.** 38.1°   **9.** 17°30′   **11.** 49°42′
13. $\sin \theta = \frac{7}{25}$, $\cos \theta = \frac{24}{25}$, $\tan \theta = \frac{7}{24}$, $\cot \theta = \frac{24}{7}$, $\sec \theta = \frac{25}{24}$, $\csc \theta = \frac{25}{7}$
15. $\sin \theta = \frac{1}{\sqrt{2}}$, $\cos \theta = \frac{1}{\sqrt{2}}$, $\tan \theta = 1$, $\cot \theta = 1$, $\sec \theta = \sqrt{2}$, $\csc \theta = \sqrt{2}$ **17.** 0.923, 2.40   **19.** 0.447, 1.12
21. 0.952   **23.** 1.853   **25.** 1.05   **27.** 8.074   **29.** 18.2°   **31.** 57.57°   **33.** 12.25°   **35.** 66.8°
37. $a = 1.83$, $B = 73.0°$, $c = 6.27$ **39.** $A = 51.5°$, $B = 38.5°$, $c = 104$ **41.** $B = 52.5°$, $b = 15.6$, $c = 19.7$
43. $A = 31.61°$, $a = 4.006$, $B = 58.39°$ **45.** $a = 0.6292$, $B = 40.33°$, $b = 0.5341$
47. $A = 48.813°$, $B = 41.187°$, $b = 10.196$ **49.** 44.4 N   **51.** 679.2 m^2 **53.** 12.0°   **55.** 4.92 km   **57.** 56%
59. 1.4 cm   **61.** 4.43 m   **63.** 0.184 km   **65.** 10.2 cm   **67.** $d = 880 \cot 2.2° \cong 440 \cot 1.1° = 22\,900$ m
69. 4810 m   **71.** 1.83 km   **73.** 464 m   **75.** 73.3 cm

Exercises 4-1, page 111

1. Yes, no   **3.** Yes, yes   **5.** $-1, -16$ **7.** $\frac{1}{4}, -0.6$ **9.** Yes   **11.** No   **13.** No   **15.** Yes
17. Yes   **19.** No

Exercises 4-2, page 116

1. 4   **3.** -5 **5.** $\frac{2}{7}$ **7.** $\frac{1}{2}$
9.

11.

13.

15.

17. $m = -2, b = 1$

(0, 1) 1 −2

19. $m = 1, b = 4$

(0, 4) 1 1

21. $m = \frac{5}{2}, b = -2$

5 (0, −2) 2

23. $m = -\frac{1}{3}, b = 1$

(0, 1) 3 −1

25.

(0, 2) (4, 0)

27.

(3, 0) (0, −4)

29.

(0, 6) (−2, 0)

31.

(0, 3) (3, 0)

33.

d 3.2 1.2 l 10

35.

F_2 50 60 F_1

Exercises 4-3, page 119

1. $x = 3.0, y = 1.0$ **3.** $x = 3.0, y = 0.0$ **5.** $x = 2.2, y = -0.3$ **7.** $x = -0.9, y = -2.3$
9. $x = 6.2, y = -0.5$ **11.** $x = 0.0, y = 3.0$ **13.** $x = -14.0, y = -5.0$ **15.** $x = 4.0, y = 7.5$
17. $x = -3.6, y = -1.4$ **19.** $x = 1.1, y = 0.8$ **21.** Dependent   **23.** $x = -1.2, y = -3.6$
25. $x = 1.5, y = 4.5$ **27.** Inconsistent   **29.** 50 N, 47 N   **31.** 8 s, 4 s

Exercises 4-4, page 125

1. $x = 1, y = -2$ **3.** $x = 7, y = 3$ **5.** $x = -1, y = -4$ **7.** $x = \frac{1}{2}, y = 2$ **9.** $x = -\frac{1}{3}, y = 4$
11. $x = \frac{9}{22}, y = -\frac{16}{11}$ **13.** $x = 3, y = 1$ **15.** $x = -1, y = -2$ **17.** $x = 1, y = 2$ **19.** Inconsistent
21. $x = -\frac{14}{5}, y = -\frac{16}{5}$ **23.** $x = 2.38, y = 0.45$ **25.** $x = \frac{1}{2}, y = -4$ **27.** $x = -\frac{2}{3}, y = 0$
29. Dependent   **31.** $x = -1, y = -2$ **33.** $V_1 = 9$ V, $V_2 = 6$ V   **35.** $x = 6250$ L, $y = 3750$ L
37. $t_1 = 32$ s, $t_2 = 20$ s   **39.** 4.0×10^6 calc/s, 2.5×10^6 calc/s   **41.** 750 mL, 250 mL
43. Incorrect conclusion or error in sales figures; system of equations is inconsistent

Exercises 4-5, page 131

1. -10 **3.** 29   **5.** 32   **7.** 93   **9.** 0.9   **11.** 96   **13.** $x = 3, y = 1$ **15.** $x = -1, y = -2$
17. $x = 1, y = 2$ **19.** Inconsistent   **21.** $x = -\frac{14}{5}, y = -\frac{16}{5}$ **23.** $x = 2.38, y = 0.45$ **25.** $x = \frac{1}{2}, y = -4$
27. $x = -\frac{2}{3}, y = 0$ **29.** Dependent   **31.** $x = -1, y = -2$ **33.** $F_1 = 15$ N, $F_2 = 6$ N
35. $A = \$3200, B = \$2800$ **37.** 2.5 h, 2.1 h   **39.** 4200, 1400   **41.** 0.8 L, 1.2 L   **43.** $V = 4.5i - 3.2$

Exercises 4-6, page 137

1. $x = 2, y = -1, z = 1$ **3.** $x = 4, y = -3, z = 3$ **5.** $x = \frac{1}{2}, y = \frac{2}{3}, z = \frac{1}{6}$ **7.** $x = \frac{2}{3}, y = -\frac{1}{3}, z = 1$
9. $x = \frac{4}{15}, y = -\frac{3}{5}, z = \frac{1}{3}$ **11.** $x = -2, y = \frac{2}{3}, z = \frac{1}{3}$ **13.** $x = \frac{3}{4}, y = 1, z = -\frac{1}{2}$
15. $r = 0, s = 0, t = 0, u = -1$ **17.** $P = 800$ h, $M = 125$ h, $I = 225$ h
19. $F_1 = 9.43$ N, $F_2 = 8.33$ N, $F_3 = 1.67$ N   **21.** $A = 22.5°, B = 45.0°, C = 112.5°$ **23.** 70 kg, 100 kg, 30 kg
25. Unlimited; $x = -10, y = -6, z = 0$ **27.** No solution

Exercises 4-7, page 143

1. 122   **3.** 651   **5.** -439 **7.** 202   **9.** 128   **11.** 0.128   **13.** $x = -1, y = 2, z = 0$
15. $x = 2, y = -1, z = 1$ **17.** $x = 4, y = -3, z = 3$ **19.** $x = \frac{1}{2}, y = \frac{2}{3}, z = \frac{1}{6}$ **21.** $x = \frac{2}{3}, y = -\frac{1}{3}, z = 1$
23. $x = \frac{4}{15}, y = -\frac{3}{5}, z = \frac{1}{3}$ **25.** $x = -2, y = \frac{2}{3}, z = \frac{1}{3}$ **27.** $x = \frac{3}{4}, y = 1, z = -\frac{1}{2}$
29. $A = 125$ N, $B = 60$ N, $F = 75$ N   **31.** 45 km/h, 540 km/h, 30 km/h

Review Exercises for Chapter 4, page 145

1. -17 **3.** -1485 **5.** -4 **7.** $\frac{2}{7}$
9. $m = -2$, $b = 4$ **11.** $m = 4$, $b = -\frac{5}{2}$

13. $x = 2.0$, $y = 0.0$ **15.** $x = 2.2$, $y = 2.7$ **17.** $x = 1.5$, $y = -1.9$ **19.** $x = 1.5$, $y = 0.4$
21. $x = 1$, $y = 2$ **23.** $x = \frac{1}{2}$, $y = -2$ **25.** $x = -\frac{1}{3}$, $y = \frac{7}{4}$ **27.** $x = \frac{11}{19}$, $y = -\frac{26}{19}$
29. $x = -\frac{6}{19}$, $y = \frac{36}{19}$ **31.** $x = 1.10$, $y = 0.54$ **33.** $x = 1$, $y = 2$ **35.** $x = \frac{1}{2}$, $y = -2$
37. $x = -\frac{1}{3}$, $y = \frac{7}{4}$ **39.** $x = \frac{11}{19}$, $y = -\frac{26}{19}$ **41.** $x = -\frac{6}{19}$, $y = \frac{36}{19}$ **43.** $x = 1.10$, $y = 0.54$
45. -115 **47.** 230.08 **49.** $x = 2$, $y = -1$, $z = 1$ **51.** $x = \frac{2}{3}$, $y = -\frac{1}{2}$, $z = 0$ **53.** $r = 3$, $s = -1$, $t = \frac{3}{2}$
55. $x = -\frac{1}{2}$, $y = \frac{1}{2}$, $z = 3$ **57.** $x = 2$, $y = -1$, $z = 1$ **59.** $x = \frac{2}{3}$, $y = -\frac{1}{2}$, $z = 0$ **61.** $r = 3$, $s = -1$, $t = \frac{3}{2}$
63. $x = -\frac{1}{2}$, $y = \frac{1}{2}$, $z = 3$ **65.** $x = \frac{8}{3}$, $y = -8$ **67.** $x = 1$, $y = 3$ **69.** -6 **71.** $-\frac{4}{3}$
73. $F_1 = 232$ N, $F_2 = 24$ N, $F_3 = 201$ N   **75.** $a = 440$ m·°C, $b = 9.6$°C   **77.** 22 800 km/h, 1400 km/h
79. $R_1 = 0.5\ \Omega$, $R_2 = 1.5\ \Omega$ **81.** $L = 10$ N, $w = 40$ N   **83.** 425 TVs, 475 VCRs, 850 CDs

Exercises 5-1, page 152

1. $40x - 40y$ **3.** $2x^3 - 8x^2$ **5.** $y^2 - 36$ **7.** $9v^2 - 4$ **9.** $16x^2 - 25y^2$ **11.** $144 - 25a^2b^2$
13. $25f^2 + 40f + 16$ **15.** $4x^2 + 28x + 49$ **17.** $x^2 - 2x + 1$ **19.** $16a^2 + 56axy + 49x^2y^2$
21. $16x^2 - 16xy + 4y^2$ **23.** $36s^2 - 12st + t^2$ **25.** $x^2 + 6x + 5$ **27.** $c^2 + 9c + 18$ **29.** $6x^2 + 13x - 5$
31. $20x^2 - 21x - 5$ **33.** $20v^2 + 13v - 15$ **35.** $6x^2 - 13xy - 63y^2$ **37.** $2x^2 - 8$ **39.** $8a^3 - 2a$
41. $6ax^2 + 24abx + 24ab^2$ **43.** $20n^4 + 100n^3 + 125n^2$ **45.** $16a^3 - 48a^2 + 36a$
47. $x^2 + y^2 + 2xy + 2x + 2y + 1$ **49.** $x^2 + y^2 + 2xy - 6x - 6y + 9$ **51.** $125 - 75t + 15t^2 - t^3$
53. $8x^3 + 60x^2t + 150xt^2 + 125t^3$ **55.** $x^2 + 2xy + y^2 - 1$ **57.** $x^3 + 8$ **59.** $64 - 27x^3$

61. $P_0P_1c + P_1G$ **63.** $4p^2 + 8pDA + 4D^2A^2$ **65.** $\frac{1}{2}\pi R^2 - \frac{1}{2}\pi r^2$ **67.** $\frac{L}{6}x^3 - \frac{L}{2}ax^2 + \frac{L}{2}a^2x - \frac{L}{6}a^3$

Exercises 5-2, page 156

1. $6(x + y)$ **3.** $5(a - 1)$ **5.** $3x(x - 3)$ **7.** $7b(by - 4)$ **9.** $6n(2n + 1)$ **11.** $2(x + 2y - 4z)$
13. $3ab(b - 2 + 4b^2)$ **15.** $4pq(3q - 2 - 7q^2)$ **17.** $2(a^2 - b^2 + 2c^2 - 3d^2)$ **19.** $(x + 2)(x - 2)$

21. $(10 + y)(10 - y)$ **23.** $(6a + 1)(6a - 1)$ **25.** $(9s + 5t)(9s - 5t)$ **27.** $(12n + 13p^2)(12n - 13p^2)$
29. $(x + y + 3)(x + y - 3)$ **31.** $2(x + 2)(x - 2)$ **33.** $3(x + 3z)(x - 3z)$ **35.** $2(a - 1)(a - 5)$

37. $(x^2 + 4)(x + 2)(x - 2)$ **39.** $(x^4 + 1)(x^2 + 1)(x + 1)(x - 1)$ **41.** $\dfrac{3 + b}{2 - b}$ **43.** $\dfrac{3}{2(t - 1)}$

45. $(3 + b)(x - y)$ **47.** $(a - b)(a + x)$ **49.** $(x + 2)(x - 2)(x + 3)$ **51.** $(x - y)(x + y + 1)$

53. $Rv(1 + v + v^2)$ **55.** $a(D_1 + D_2)(D_1 - D_2)$ **57.** $Pb(L + b)(L - b)$ **59.** $\dfrac{5Y}{9S - 3Y}$

Exercises 5-3, page 162

1. $(x + 1)(x + 4)$ **3.** $(s - 7)(s + 6)$ **5.** $(t + 8)(t - 3)$ **7.** $(x + 1)^2$ **9.** $(x - 2y)^2$ **11.** $(3x + 1)(x - 2)$
13. $(3y + 1)(y - 3)$ **15.** $(2s + 11)(s + 1)$ **17.** $(3f - 1)(f - 5)$ **19.** $(2t - 3)(t + 5)$ **21.** $(3t - 4u)(t - u)$
23. $(4x - 7)(x + 1)$ **25.** $(9x - 2y)(x + y)$ **27.** $(2m + 5)^2$ **29.** $(2x - 3)^2$ **31.** $(3t - 4)(3t - 1)$
33. $(8b - 1)(b + 4)$ **35.** $(4p - q)(p - 6q)$ **37.** $(12x - y)(x + 4y)$ **39.** $2(x - 1)(x - 6)$
41. $2(2x - 1)(x + 4)$ **43.** $ax(x + 6a)(x - 2a)$ **45.** $(a + b + 2)(a + b - 2)$ **47.** $(5a + 5x + y)(5a - 5x - y)$
49. $(x + 1)^3$ **51.** $(2x + 1)(4x^2 - 2x + 1)$ **53.** $4(s + 1)(s + 3)$ **55.** $100(2n + 3)(n - 12)$
57. $wx^2(x - 2L)(x - 3L)$ **59.** $Ad(3u - v)(u - v)$

Exercises 5-4, page 166

1. $\dfrac{14}{21}$ **3.** $\dfrac{2ax^2}{2xy}$ **5.** $\dfrac{2x - 4}{x^2 + x - 6}$ **7.** $\dfrac{ax^2 - ay^2}{x^2 - xy - 2y^2}$ **9.** $\dfrac{7}{11}$ **11.** $\dfrac{2xy}{4y^2}$ **13.** $\dfrac{2}{x + 1}$ **15.** $\dfrac{x - 5}{2x - 1}$

17. $\dfrac{1}{4}$ **19.** $\dfrac{3}{4}x$ **21.** $\dfrac{1}{5a}$ **23.** $\dfrac{3a - 2b}{2a - b}$ **25.** $\dfrac{4x^2 + 1}{(2x + 1)(2x - 1)}$ (cannot be reduced)   **27.** $3x$ **29.** $\dfrac{1}{2y^2}$

72

31. $\dfrac{x-4}{x+4}$ **33.** $\dfrac{2x-1}{x+8}$ **35.** $\dfrac{5x+4}{x(x+3)}$ **37.** $(x^2+4)(x-2)$ **39.** $\dfrac{x^2y^2(y+x)}{y-x}$ **41.** $\dfrac{x+3}{x-3}$ **43.** $-\dfrac{1}{2}$

45. $-\dfrac{2x-1}{x}$ **47.** $\dfrac{(x+5)(x-3)}{(5-x)(x+3)}$ **49.** $\dfrac{x^2-xy+y^2}{2}$ **51.** $\dfrac{(x+1)^2}{x^2-x+1}$ **53.** (a)   **55.** (a)   **57.** $\dfrac{4r}{3\pi}$

59. $\dfrac{E^2(R-r)}{(R+r)^3}$

Exercises 5-5, page 170

1. $\dfrac{3}{28}$ **3.** $6xy$ **5.** $\dfrac{7}{18}$ **7.** $\dfrac{xy^2}{bz^2}$ **9.** $4t$ **11.** $3(u+v)(u-v)$ **13.** $\dfrac{10}{3(a+4)}$ **15.** $\dfrac{x-3}{x(x+3)}$ **17.** $\dfrac{3x}{5a}$

19. $\dfrac{(x+1)(x-1)(x-4)}{4(x+2)}$ **21.** $\dfrac{x^2}{a+x}$ **23.** $\dfrac{15}{4}$ **25.** $\dfrac{3}{4x+3}$ **27.** $\dfrac{7x-1}{3x+5}$ **29.** $\dfrac{7x^4}{3a^4}$

31. $\dfrac{4t(2t-1)(t+5)}{(2t+1)^2}$ **33.** $\dfrac{1}{2}(x+y)$ **35.** $(x+y)(3p+7q)$ **37.** $\dfrac{na^2}{V(1-a)}$ **39.** $\dfrac{\pi}{2}$

Exercises 5-6, page 176

1. $\dfrac{9}{5}$ **3.** $\dfrac{8}{x}$ **5.** $\dfrac{5}{4}$ **7.** $\dfrac{3+7ax}{4x}$ **9.** $\dfrac{ax-b}{x^2}$ **11.** $\dfrac{30+ax^2}{25x^3}$ **13.** $\dfrac{14-a^2}{10a}$

15. $\dfrac{-x^2+4x+xy+y-2}{xy}$ **17.** $\dfrac{7}{2(2x-1)}$ **19.** $\dfrac{5-3x}{2x(x+1)}$ **21.** $\dfrac{-3}{4(s-3)}$ **23.** $\dfrac{7x+6}{3(x+3)(x-3)}$

25. $\dfrac{2x-5}{(x-4)^2}$ **27.** $\dfrac{x+27}{(x-5)(x+5)(x-6)}$ **29.** $\dfrac{9x^2+x-2}{(3x-1)(x-4)}$ **31.** $\dfrac{13t^2+27t}{(t-3)(t+2)(t+3)^2}$ **33.** $\dfrac{x+1}{x-1}$

35. $-\dfrac{(x+1)(x-1)(2x+1)}{x^2(x+2)}$ **37.** $-\dfrac{(3x+4)(x-1)}{2x}$ **39.** $\dfrac{2s}{r-s}$ **41.** $\dfrac{h}{(x+1)(x+h+1)}$ **43.** $\dfrac{-2hx-h^2}{x^2(x+h)^2}$

45. $\dfrac{y^2-rx+r^2}{r^2}$ **47.** $\dfrac{2a-1}{a^2}$ **49.** $\dfrac{3(H-H_0)}{4\pi H}$ **51.** $\dfrac{n(2n+1)}{2(n+2)(n-1)}$

Exercises 5-7, page 180

1. 4 **3.** -3 **5.** $\dfrac{7}{2}$ **7.** $\dfrac{16}{21}$ **9.** -9 **11.** $-\dfrac{2}{13}$ **13.** $\dfrac{5}{3}$ **15.** -2 **17.** $\dfrac{3}{4}$ **19.** 6 **21.** -5

23. $-\dfrac{7}{8}$ **25.** No solution   **27.** $\dfrac{2}{3}$ **29.** $\dfrac{3b}{1-2b}$ **31.** $\dfrac{(2b-1)(b+6)}{2(b-1)}$ **33.** $\dfrac{n_1V-nV}{n_1}$ **35.** $\dfrac{V_rA-V_0}{V_0A}$

37. $\dfrac{jX}{1-g_mz}$ **39.** $\dfrac{PV^3-bPV^2+aV-ab}{RV^2}$ **41.** $\dfrac{kA_1A_2R-A_1L_2}{A_2}$ **43.** $\dfrac{fnR_2-fR_2}{R_2+f-fn}$ **45.** 2.4 h

47. 3.2 min   **49.** 80 km/h   **51.** 2.5 L

Review Exercises for Chapter 5, page 182

1. $12ax+15a^2$ **3.** $4a^2-49b^2$ **5.** $4a^2+4a+1$ **7.** $b^2+3b-28$ **9.** $2x^2-13x-45$
11. $16c^2+6cd-d^2$ **13.** $3(s+3t)$ **15.** $a^2(x^2+1)$ **17.** $(x+12)(x-12)$ **19.** $(4x+8+t^2)(4x+8-t^2)$
21. $(3t-1)^2$ **23.** $(5t+1)^2$ **25.** $(x+8)(x-7)$ **27.** $(t-9)(t+4)$ **29.** $(2x-9)(x+4)$
31. $(2x+5)(2x-7)$ **33.** $(5b-1)(2b+5)$ **35.** $4(x+4y)(x-4y)$ **37.** $(x+3)^3$

39. $(2x+3)(4x^2-6x+9)$ **41.** $(a-3)(b^2+1)$ **43.** $(x+5)(n-x+5)$ **45.** $\dfrac{16x^2}{3a^2}$ **47.** $\dfrac{3x+1}{2x-1}$

49. $\dfrac{16}{5x(x-y)}$ **51.** $\dfrac{6}{5-x}$ **53.** $\dfrac{x+2}{2x(7x-1)}$ **55.** $\dfrac{1}{x-1}$ **57.** $\dfrac{16x-15}{36x^2}$ **59.** $\dfrac{5y+6}{2xy}$ **61.** $\dfrac{-2(2a+3)}{a(a+2)}$

63. $\dfrac{2x^2-x+1}{x(x+3)(x-1)}$ **65.** $\dfrac{12x^2-7x-4}{2(x-1)(x+1)(4x-1)}$ **67.** $\dfrac{x^3+6x^2-2x+2}{x(x-1)(x+3)}$ **69.** 2 **71.** $\dfrac{7}{2c+4}$

73. $-\dfrac{(a-1)^2}{2a}$ **75.** 6 **77.** $\dfrac{1}{4}[(x+y)^2-(x-y)^2]=\dfrac{1}{4}(x^2+2xy+y^2-x^2+2xy-y^2)=\dfrac{1}{4}(4xy)=xy$

79. $2zS^2+2zS$ **81.** $\pi\ell(r_1+r_2)(r_1-r_2)$ **83.** $(t+1)(1-t)$ **85.** $W^4-4L^4+4k^2L^2$

87. $4(3x^2+12x+16)$ **89.** $\dfrac{12\pi^2wv^2D}{gn^2t}$ **91.** $\dfrac{60t+9t^2+2t^3}{6}$ **93.** $\dfrac{120-60d^2+5d^4-d^6}{120}$

95. $\dfrac{2\pi^2 CN + 2\pi^2 Cn + N^2 - 2nN + n^2}{4\pi^2 C}$ **97.** $\dfrac{4r^3 - 3ar^2 - a^3}{4r^3}$ **99.** $\dfrac{q_2 D - fd}{D + d}$ **101.** $\dfrac{RHw}{w - RH}$

103. $\dfrac{-mb^2 s^2 - kL^2}{b^2 s}$ **105.** 3.4 h   **107.** 15 s   **109.** 11.3   **111.** 18 Ω

Exercises 6-1, page 189

1. $a = 1, b = -8, c = 5$ **3.** $a = 1, b = -2, c = -4$ **5.** Not quadratic   **7.** $a = 1, b = -1, c = 0$
9. $2, -2$ **11.** $\frac{3}{2}, -\frac{3}{2}$ **13.** $-1, 9$ **15.** $3, 4$ **17.** $0, -2$ **19.** $\frac{1}{3}, -\frac{1}{3}$ **21.** $\frac{1}{3}, 4$ **23.** $-4, -4$
25. $\frac{2}{3}, \frac{3}{2}$ **27.** $\frac{1}{2}, -\frac{3}{2}$ **29.** $2, -1$ **31.** $2b, -2b$ **33.** $0, \frac{5}{2}$ **35.** $\frac{5}{2}, -\frac{9}{2}$ **37.** $0, -2$ **39.** $b - a, -b - a$
41. 10 Pa   **43.** 2 A.M., 10 A.M.   **45.** $\frac{3}{2}, 4$ **47.** $-2, \frac{1}{2}$ **49.** 3 N/cm, 6 N/cm   **51.** 800 km/h

Exercises 6-2, page 193

1. $-5, 5$ **3.** $-\sqrt{7}, \sqrt{7}$ **5.** $-3, 7$ **7.** $-3 \pm \sqrt{7}$ **9.** $2, -4$ **11.** $-2, -1$ **13.** $2 \pm \sqrt{2}$
15. $-5, 3$ **17.** $-3, \frac{1}{2}$ **19.** $\frac{1}{6}(3 \pm \sqrt{33})$ **21.** $\frac{1}{4}(1 \pm \sqrt{17})$ **23.** $-b \pm \sqrt{b^2 - c}$

Exercises 6-3, page 197

1. $2, -4$ **3.** $-2, -1$ **5.** $2 \pm \sqrt{2}$ **7.** $-5, 3$ **9.** $-3, \frac{1}{2}$ **11.** $\frac{1}{6}(3 \pm \sqrt{33})$ **13.** $\frac{1}{4}(1 \pm \sqrt{17})$
15. $-\frac{8}{3}, \frac{5}{6}$ **17.** $\frac{1}{2}(-5 \pm \sqrt{-5})$ **19.** $\frac{1}{6}(1 \pm \sqrt{109})$ **21.** $\frac{3}{2}, -\frac{3}{2}$ **23.** $\frac{3}{4}, -\frac{5}{8}$ **25.** $-0.54, 0.74$
27. $-0.26, 2.43$ **29.** $-c \pm \sqrt{c^2 + 1}$ **31.** $\dfrac{b + 1 \pm \sqrt{-3b^2 + 2b + 4b^2 a + 1}}{2b^2}$ **33.** 4.376 cm
35. $\dfrac{-R \pm \sqrt{R^2 - 4L/C}}{2L}$ **37.** 4.5 m, 6.5 m   **39.** 61.7 m by 32.4 m or 21.6 m by 92.6 m

Exercises 6-4, page 202

1.

3.

5.
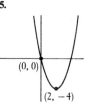

7.
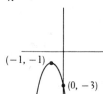

9.
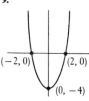

11.

13.

15.

17. $-1.2, 1.2$ **19.** $0.5, 3.1$ **21.** No real roots   **23.** $-2.4, 1.2$
25.
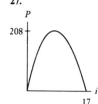
27.

29. (a) After 19 s,   **31.** 1.1 m
   (b) 460 m,
   (c) 1.2 s, 17.2 s
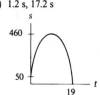

1. $-4, 1$ **3.** $2, 8$ **5.** $\frac{1}{3}, -4$ **7.** $\frac{1}{2}, \frac{5}{3}$ **9.** $0, \frac{25}{6}$ **11.** $-\frac{3}{2}, \frac{7}{2}$ **13.** $-10, 11$ **15.** $-1 \pm \sqrt{6}$
17. $-4, \frac{9}{2}$ **19.** $\frac{1}{8}(3 \pm \sqrt{41})$ **21.** $\frac{1}{42}(-23 \pm \sqrt{-4091})$ **23.** $\frac{1}{6}(-2 \pm \sqrt{58})$ **25.** $-2 \pm 2\sqrt{2}$
27. $\frac{1}{3}(-4 \pm \sqrt{10})$ **29.** $-1, \frac{5}{4}$ **31.** $\frac{1}{4}(-3 \pm \sqrt{-47})$ **33.** $\dfrac{-1 \pm \sqrt{-1}}{a}$ **35.** $\dfrac{-3 \pm \sqrt{9 + 4a^3}}{2a}$
37. $-5, 6$ **39.** $\frac{1}{4}(1 \pm \sqrt{33})$ **41.** $3 \pm \sqrt{7}$ **43.** 0 (3 is not a solution)

45.

47.

49. $-1.7, 1.2$ **51.** No real roots
53. 1.2 cm, 4.0 cm   **55.** 0.6 s, 2.2 s
57. 8000   **59.** $\dfrac{-k \pm \sqrt{k^2 + 4ck}}{2c}$

61.

63. 61 m, 81 m   **65.** 3.06 mm   **67.** 1.1 m   **69.** 25
71.

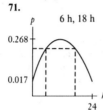

6 h, 18 h

Exercises 7-1, page 207

1. $+, -, -$ **3.** $+, +, -$ **5.** $+, +, +$ **7.** $+, -, +$
9. $\sin \theta = \dfrac{1}{\sqrt{5}}$, $\cos \theta = \dfrac{2}{\sqrt{5}}$, $\tan \theta = \frac{1}{2}$, $\cot \theta = 2$, $\sec \theta = \frac{1}{2}\sqrt{5}$, $\csc \theta = \sqrt{5}$

11. $\sin \theta = -\dfrac{3}{\sqrt{13}}$, $\cos \theta = -\dfrac{2}{\sqrt{13}}$, $\tan \theta = \frac{3}{2}$, $\cot \theta = \frac{2}{3}$, $\sec \theta = -\frac{1}{2}\sqrt{13}$, $\csc \theta = -\frac{1}{3}\sqrt{13}$

13. $\sin \theta = \frac{12}{13}$, $\cos \theta = -\frac{5}{13}$, $\tan \theta = -\frac{12}{5}$, $\cot \theta = -\frac{5}{12}$, $\sec \theta = -\frac{13}{5}$, $\csc \theta = \frac{13}{12}$

15. $\sin \theta = -\dfrac{2}{\sqrt{29}}$, $\cos \theta = \dfrac{5}{\sqrt{29}}$, $\tan \theta = -\frac{2}{5}$, $\cot \theta = -\frac{5}{2}$, $\sec \theta = \frac{1}{5}\sqrt{29}$, $\csc \theta = -\frac{1}{2}\sqrt{29}$ **17.** II   **19.** II
21. IV   **23.** III

Exercises 7-2, page 214

1. $\sin 20°$; $-\cos 40°$ **3.** $-\tan 75°$, $-\csc 58°$ **5.** $-\sin 57°$; $-\cot 6°$ **7.** $\cos 40°$; $-\tan 40°$
9. $-\sin 15° = -0.26$ **11.** $-\cos 73.7° = -0.281$ **13.** $\tan 39.15° = 0.8141$ **15.** $\sec 31.67° = 1.175$
17. -0.523 **19.** -0.7620 **21.** -0.34 **23.** -3.910 **25.** $237.99°, 302.01°$ **27.** $66.40°, 293.60°$
29. $15.8°, 195.8°$ **31.** $102.0°, 282.0°$ **33.** $119.5°$ **35.** $263°$ **37.** $306.21°$ **39.** $299.24°$ **41.** -0.7003
43. -0.777 **45.** $<$ **47.** $=$ **49.** 0.0183 A   **51.** 16.7 cm
53. $\cos(-\theta) = \dfrac{x}{r}$, $\tan(-\theta) = \dfrac{-y}{x}$, $\cot(-\theta) = \dfrac{x}{-y}$, $\sec(-\theta) = \dfrac{r}{x}$, $\csc(-\theta) = \dfrac{r}{-y}$ **55.** (a) 5.7, (b) -1.4

Exercises 7-3, page 219

1. $\dfrac{\pi}{12}, \dfrac{5\pi}{6}$ **3.** $\dfrac{5\pi}{12}, \dfrac{11\pi}{6}$ **5.** $\dfrac{7\pi}{6}, \dfrac{3\pi}{2}$ **7.** $\dfrac{8\pi}{9}, \dfrac{13\pi}{9}$ **9.** $72°, 270°$ **11.** $10°, 315°$ **13.** $170°, 300°$
15. $15°, 27°$ **17.** 0.401 **19.** 4.40 **21.** 5.821 **23.** 3.115 **25.** $43.0°$ **27.** $195.2°$ **29.** $140°$
31. $940.8°$ **33.** 0.7071 **35.** 3.732 **37.** -0.8660 **39.** -8.327 **41.** 0.9056 **43.** -0.89
45. -0.48 **47.** -0.15 **49.** $0.3141, 2.827$ **51.** $2.932, 6.074$ **53.** $0.8309, 5.452$ **55.** $2.442, 3.841$
57. 0.030 N·m   **59.** 2400 m

Exercises 7-4, page 224

1. 346 m   **3.** 5570 cm^2 **5.** 0.382 = 21.9°   **7.** 627 m^2 **9.** 0.52 rad/s   **11.** 34.73 m^2 **13.** 0.704 m
15. 22.6 m^2 **17.** 369 m^3 **19.** 0.4 km   **21.** 8.12 r/min   **23.** 150 m/s   **25.** 9.41 m
27. 35.9 m/min   **29.** 0.433 rad/s   **31.** 7540 m/min   **33.** 209 rad   **35.** 14.9 m^3
37. 4.848 × 10^{-6} (all three values)   **39.** 1.15 × 10^8 km

Review Exercises for Chapter 7, page 227

1. $\sin \theta = \frac{4}{5}$, $\cos \theta = \frac{3}{5}$, $\tan \theta = \frac{4}{3}$, $\cot \theta = \frac{3}{4}$, $\sec \theta = \frac{5}{3}$, $\csc \theta = \frac{5}{4}$

3. $\sin \theta = -\dfrac{2}{\sqrt{53}}$, $\cos \theta = \dfrac{7}{\sqrt{53}}$, $\tan \theta = -\frac{2}{7}$, $\cot \theta = -\frac{7}{2}$, $\sec \theta = \dfrac{\sqrt{53}}{7}$, $\csc \theta = -\dfrac{\sqrt{53}}{2}$

5. $-\cos 48°$, $\tan 14°$ **7.** $-\sin 71°$, $\sec 15°$ **9.** $\dfrac{2\pi}{9}$, $\dfrac{17\pi}{20}$ **11.** $\dfrac{4\pi}{15}$, $\dfrac{9\pi}{8}$ **13.** 252°, 130°   **15.** 12°, 330°

17. 32.1°   **19.** 206.7°   **21.** 1.78   **23.** 0.3534   **25.** 4.5736   **27.** 2.377   **29.** −0.415   **31.** −0.47
33. −1.080   **35.** −0.4264   **37.** −1.64   **39.** 4.140   **41.** −0.5878   **43.** −0.8660   **45.** 0.5569
47. 1.197   **49.** 10.30°, 190.30°   **51.** 118.23°, 241.77°   **53.** 0.5759, 5.707   **55.** 4.187, 5.238   **57.** 223.76°
59. 246.78°   **61.** 0.0562 W   **63.** 0.800 = 45.8°   **65.** 18.98 cm   **67.** 2.70 m^2 **69.** 4710 cm/s
71. 1.81 × 10^6 cm/s   **73.** 51.27 m   **75.** 3.58 × 10^5 km

Exercises 8-1, page 234

1. (a) Vector: magnitude and direction are specified; (b) scalar: only magnitude is specified
3. (a) Vector: magnitude and direction are specified; (b) scalar: only magnitude is specified
5.   **7.**   **9.**   **11.**   **13.**   **15.**

17.   **19.**   **21.**   **23.**   **25.**   **27.**

29.   **31.**   **33.**   **35.**

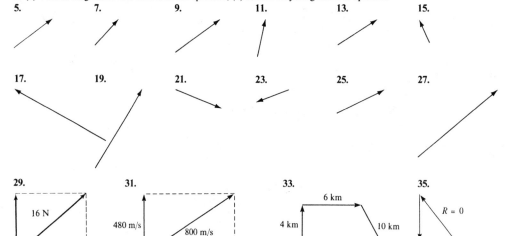

Exercises 8-2, page 237

1. 662, 352   **3.** −349, −664   **5.** 3.22, 7.97   **7.** −62.9, 44.1   **9.** 2.08, −8.80   **11.** −2.53, −0.788
13. −0.8088, 0.3296   **15.** 88 920, 12 240   **17.** 23.9 km/h, 7.43 km/h   **19.** 52.2 N, 17.4 N
21. 115 km, 88.3 km   **23.** 29 830 km/h, −1329 km/h

Exercises 8-3, page 242

1. $R = 24.2$, $\theta = 52.6°$ with A **3.** $R = 7.781$, $\theta = 66.63°$ with A **5.** $R = 10.0$, $\theta = 58.8°$
7. $R = 2.74$, $\theta = 111.0°$ **9.** $R = 2130$, $\theta = 107.7°$ **11.** $R = 1.426$, $\theta = 299.12°$ **13.** $R = 29.2$, $\theta = 10.8°$
15. $R = 47.0$, $\theta = 101.0°$ **17.** $R = 27.27$, $\theta = 33.14°$ **19.** $R = 12.735$, $\theta = 25.216°$ **21.** $R = 50.2$, $\theta = 50.3°$
23. $R = 235$, $\theta = 121.7°$

Exercises 8-4, page 245

1. 6.60 N, 29.5° from 5.75-N force   **3.** 11.4 kN, 44.2° above horizontal   **5.** 3070 m, 17.8° S of W
7. 229.4 m, 72.82° N of E   **9.** 25.3 km/h, 29.6° S of E   **11.** 175 N, 9.3° above horizontal
13. 540 km/h, 6.2° from direction of plane   **15.** 29 180 km/h, 0.03° from direction of shuttle
17. 1810 m/min^2, $\phi = 89.6°$ **19.** 138 km, 65.0° N of E
21. 79.0 m/s, 11.3° from direction of plane, 75.6° from vertical   **23.** 4.06 A/m, 11.6° with magnet

Exercises 8-5, page 252

1. $b = 38.1$, $C = 66.0°$, $c = 46.1$ **3.** $a = 2800$, $b = 2620$, $C = 108.0°$ **5.** $B = 12.20°$, $C = 149.57°$, $c = 7.448$
7. $a = 110.5$, $A = 149.70°$, $C = 9.57°$ **9.** $A = 125.6°$, $a = 0.0776$, $c = 0.005\,66$ **11.** $A = 99.4°$, $b = 55.1$, $c = 24.4$
13. $A = 68.01°$, $a = 5520$, $c = 5376$
15. $A_1 = 61.36°$, $C_1 = 70.51°$, $c_1 = 5.628$; $A_2 = 118.64°$, $C_2 = 13.23°$, $c_2 = 1.366$
17. $A_1 = 107.3°$, $a_1 = 5280$, $C_1 = 41.3°$; $A_2 = 9.9°$, $a_2 = 952$, $C_2 = 138.7°$ **19.** No solution   **21.** 0.803 m
23. 884 N   **25.** 406 m   **27.** 13.94 cm   **29.** 27 300 km   **31.** 77.3° with bank downstream

Exercises 8-6, page 257

1. $A = 50.3°$, $B = 75.7°$, $c = 6.31$ **3.** $A = 70.9°$, $B = 11.1°$, $c = 4750$ **5.** $A = 34.72°$, $B = 40.67°$, $C = 104.61°$
7. $A = 18.21°$, $B = 22.28°$, $C = 139.51°$ **9.** $A = 6.0°$, $B = 16.0°$, $c = 1150$ **11.** $A = 82.3°$, $b = 21.6$, $C = 11.4°$
13. $A = 36.24°$, $B = 39.09°$, $a = 97.22$ **15.** $A = 46.94°$, $B = 61.82°$, $C = 71.24°$
17. $A = 137.9°$, $B = 33.7°$, $C = 8.4°$ **19.** $b = 37$, $C = 25°$, $c = 24$ **21.** 69.4 km   **23.** 0.039 km
25. 57.3°, 141.7°   **27.** 5.54 m   **29.** 5.09 km/h   **31.** 94.7°

Review Exercises for Chapter 8, page 258

1. $A_x = 57.4$, $A_y = 30.5$ **3.** $A_x = -0.7485$, $A_y = -0.5357$ **5.** $R = 602$, $\theta = 57.1°$ with A
7. $R = 5960$, $\theta = 33.60°$ with A **9.** $R = 965$, $\theta = 8.6°$ **11.** $R = 26.12$, $\theta = 146.03°$
13. $R = 71.93$, $\theta = 336.50°$ **15.** $R = 99.42$, $\theta = 359.57°$ **17.** $b = 18.1$, $C = 64.0°$, $c = 17.5$
19. $A = 21.2°$, $b = 34.8$, $c = 51.5$ **21.** $a = 17\,340$, $b = 24\,660$, $C = 7.99°$ **23.** $A = 39.88°$, $a = 51.94$, $C = 30.03°$
25. $A_1 = 54.8°$, $a_1 = 12.7$, $B_1 = 68.6°$; $A_2 = 12.0°$, $a_2 = 3.24$, $B_2 = 111.4°$ **27.** $A = 32.3°$, $b = 267$, $C = 17.7°$
29. $A = 148.7°$, $B = 9.3°$, $c = 5.66$ **31.** $a = 1782$, $b = 1920$, $C = 16.00°$ **33.** $A = 37°$, $B = 25°$, $C = 118°$
35. $A = 20.6°$, $B = 35.6°$, $C = 123.8°$ **37.** -155.7 N, 81.14 N   **39.** 630 m/s   **41.** 14.9 mN
43. 3.67 m/s^2 **45.** 2.30 m, 2.49 m   **47.** 52 700 km   **49.** 2.65 km   **51.** 299 km   **53.** 808 N, 36.4° N of E
55. 1270 m or 1680 m (ambiguous)

Exercises 9-1, page 264

1. 0, -0.7, -1, -0.7, 0, 0.7, 1, 0.7, 0, -0.7, -1, -0.7, 0, 0.7, 1, 0.7, 0

3. -3, -2.1, 0, 2.1, 3, 2.1, 0, -2.1, -3, -2.1, 0, 2.1, 3, 2.1, 0, -2.1, -3

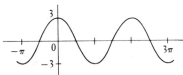

5. 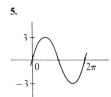   **7.** 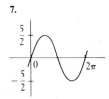   **9.** 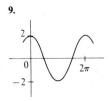   **11.**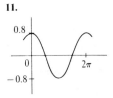

13. 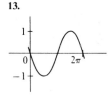   **15.** 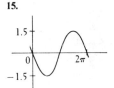   **17.** 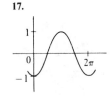   **19.**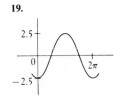

21. 0, 0.84, 0.91, 0.14, −0.76, −0.96, −0.28, 0.66

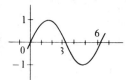

23. 1, 0.54, −0.42, −0.99, −0.65, 0.28, 0.96, 0.75

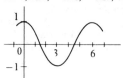

Exercises 9-2, page 268

1. $\dfrac{\pi}{3}$ **3.** $\dfrac{\pi}{4}$ **5.** $\dfrac{\pi}{6}$ **7.** $\dfrac{\pi}{8}$ **9.** 1   **11.** $\dfrac{1}{2}$ **13.** 6π **15.** 3π **17.** 3   **19.** $\dfrac{2}{\pi}$

21.

23.

25.

27.

29.

31.

33.

35.

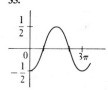

37.

39.

41. $y = \sin 6x$

43. $y = \sin \pi x$

45.

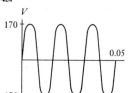

47.

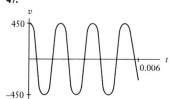

Exercises 9-3, page 271

1. $1, 2\pi, \dfrac{\pi}{6}$

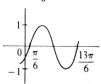

3. $1, 2\pi, -\dfrac{\pi}{6}$

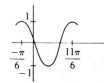

5. $2, \pi, -\dfrac{\pi}{4}$

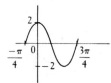

7. $1, \pi, \dfrac{\pi}{2}$

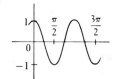

9. $\frac{1}{2}, 4\pi, \frac{\pi}{2}$

11. $3, 6\pi, -\pi$

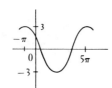

13. $1, 2, -\frac{1}{8}$

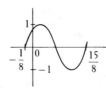

15. $\frac{3}{4}, \frac{1}{2}, \frac{1}{20}$

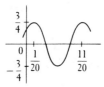

17. $0.6, 1, \frac{1}{2\pi}$

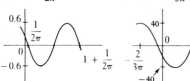

19. $40, \frac{2}{3}, -\frac{2}{3\pi}$

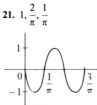

21. $1, \frac{2}{\pi}, \frac{1}{\pi}$

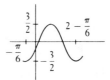

23. $\frac{3}{2}, 2, -\frac{\pi}{6}$

25.

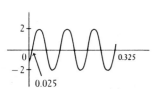

27.

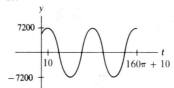

Exercises 9-4, page 275

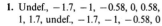

1. Undef., -1.7, -1, -0.58, 0, 0.58, 1, 1.7, undef., -1.7, -1, -0.58, 0

3. Undef., 2, 1.4, 1.2, 1, 1.2, 1.4, 2, undef., -2, -1.4, -1.2, -1

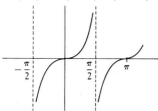

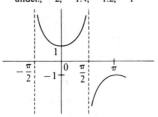

5.

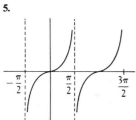

7.

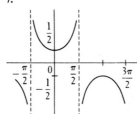

9.

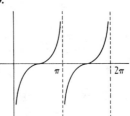

11.

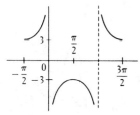

13.

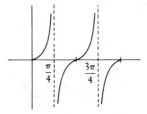

15.

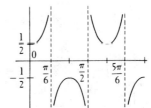

79

17.

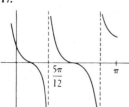

19.

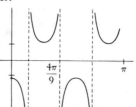

21.

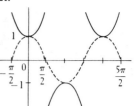

23.

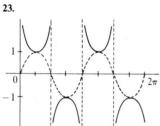

25.

27.

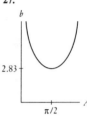

Exercises 9-5, page 279

1.

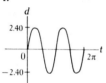

3.

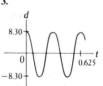

5.

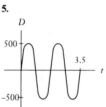

7.

9.

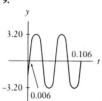

11.

13.

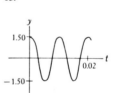

15.

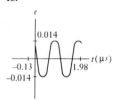

Exercises 9-6, page 284

1.

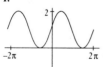

3.

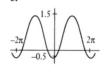

5.

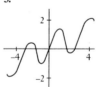

7.

9.

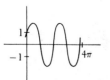

11.

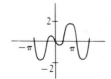

13.

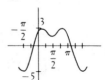

15.

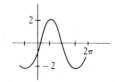

17.

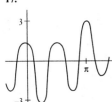

19.

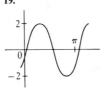

21.

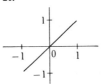

23.

25.

27.

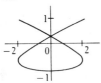

29.

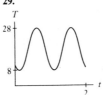

31.

33.

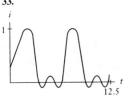

35.

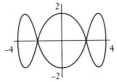

Review Exercises for Chapter 9, page 286

1.

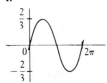

3.

5.

7.

9.

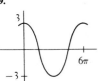

11.

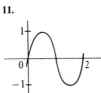

13.

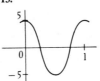

15.

17.

19.

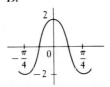

21.

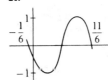

23.

25.

27.

29.

31.

33.

35.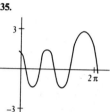

37. $y = 2 \sin\left(2x + \dfrac{\pi}{2}\right)$

39. $y = \cos\left(\dfrac{\pi}{4}x - \dfrac{3\pi}{4}\right)$

41.

43.

45.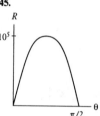

47.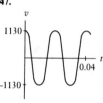

49.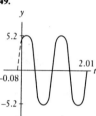

51.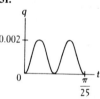

53.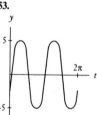

55.

57.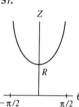

59.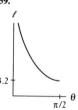

Exercises 10-1, page 292

1. x^3 **3.** $\dfrac{1}{a^4}$ **5.** $\dfrac{1}{25}$ **7.** $\dfrac{25}{4}$ **9.** $\dfrac{4a^2}{x^2}$ **11.** $\dfrac{n^2}{5a}$ **13.** 1 **15.** -7 **17.** $\dfrac{3}{x^2}$ **19.** $\dfrac{1}{7^3 a^3 x^3}$ **21.** $\dfrac{n^3}{2}$

23. $\dfrac{1}{a^3 b^6}$ **25.** $\dfrac{1}{a+b}$ **27.** $\dfrac{2x^2 + 3y^2}{x^2 y^2}$ **29.** $\dfrac{2^3}{3^5}$ **31.** $\dfrac{a^2 b^2}{3}$ **33.** $\dfrac{b^3}{432a}$ **35.** $\dfrac{4}{t^4 v^4}$ **37.** $\dfrac{x^8 - y^2}{x^4 y^2}$

39. $\dfrac{2a^6 + 16}{a^8}$ **41.** $\dfrac{10}{9}$ **43.** $\dfrac{ab}{a+b}$ **45.** $\dfrac{4n^2 - 4n + 1}{n^4}$ **47.** $\dfrac{45}{2}$ **49.** $-\dfrac{x}{y}$ **51.** $\dfrac{a^2 - ax + x^2}{ax}$

82

53. $\dfrac{t^2 + t + 2}{t^2}$　　**55.** $\dfrac{2x}{(x+1)(x-1)}$　　**57.** (a) 4^5, (b) 2^{10}　　**59.** $\left(\dfrac{a}{b}\right)^{-n} = \dfrac{1}{\left(\dfrac{a}{b}\right)^n} = \dfrac{1}{\dfrac{a^n}{b^n}} = \dfrac{b^n}{a^n} = \left(\dfrac{b}{a}\right)^n$　　**61.** $\dfrac{Pa \cdot m^3}{mol}$

63. $\dfrac{\omega^4 - 2\omega^2\omega_0^2 + \omega_0^4}{\omega^2\omega_0^2}$

Exercises 10-2, page 297

1. 5　　**3.** 3　　**5.** 16　　**7.** 10^{25}　　**9.** $\dfrac{1}{2}$　　**11.** $\dfrac{1}{16}$　　**13.** 25　　**15.** 4096　　**17.** $\dfrac{1}{110}$　　**19.** $\dfrac{6}{7}$　　**21.** $-\dfrac{1}{2}$

23. 24　　**25.** $\dfrac{39}{1000}$　　**27.** $\dfrac{3}{5}$　　**29.** 2.059　　**31.** 0.538 91　　**33.** $a^{7/6}$　　**35.** $\dfrac{1}{y^{9/10}}$　　**37.** $s^{23/12}$　　**39.** $\dfrac{1}{y^{13/12}}$

41. $2ab^2$　　**43.** $\dfrac{1}{8a^3b^{9/4}}$　　**45.** $\dfrac{4x}{(4x^2+1)^{1/2}}$　　**47.** $\dfrac{27}{64t^3}$　　**49.** $\dfrac{b^{11/10}}{2a^{1/12}}$　　**51.** $\dfrac{2}{3}x^{1/6}y^{11/12}$　　**53.** $\dfrac{x}{(x+2)^{1/2}}$

55. $\dfrac{a^2+1}{a^4}$　　**57.** $\dfrac{a+1}{a^{1/2}}$　　**59.** $\dfrac{5x^2-2x}{(2x-1)^{1/2}}$　　**61.**　　**63.**　　**65.** 0.84

67. 1.88 mA

Exercises 10-3, page 301

1. $2\sqrt{6}$　　**3.** $3\sqrt{5}$　　**5.** $xy^2\sqrt{y}$　　**7.** $qr^3\sqrt{pr}$　　**9.** $x\sqrt{5}$　　**11.** $3ac^2\sqrt{2ab}$　　**13.** $2\sqrt[3]{2}$　　**15.** $2\sqrt[3]{3}$

17. $2\sqrt[3]{a^2}$　　**19.** $2st\sqrt[4]{4r^3t}$　　**21.** 2　　**23.** $ab\sqrt[3]{b^2}$　　**25.** $\dfrac{1}{2}\sqrt{6}$　　**27.** $\dfrac{\sqrt{ab}}{b}$　　**29.** $\dfrac{1}{2}\sqrt[3]{6}$　　**31.** $\dfrac{1}{3}\sqrt[3]{27}$

33. $2\sqrt{5}$　　**35.** 2　　**37.** 200　　**39.** 2000　　**41.** $\sqrt{2a}$　　**43.** $\dfrac{1}{2}\sqrt{2}$　　**45.** $\sqrt[3]{2}$　　**47.** $\sqrt[8]{2}$　　**49.** $\dfrac{1}{6}\sqrt{6}$

51. $\dfrac{\sqrt{b(a^2+b)}}{ab}$　　**53.** $\dfrac{\sqrt{2x^2+x}}{2x+1}$　　**55.** $a+b$　　**57.** $\sqrt{4x^2-1}$　　**59.** $\dfrac{1}{2}\sqrt{4x^2+1}$　　**61.** 55.0%

63. $\dfrac{24\sqrt{EIgWL}}{WL^2}$

Exercises 10-4, page 304

1. $7\sqrt{3}$　　**3.** $\sqrt{5}-\sqrt{7}$　　**5.** $3\sqrt{5}$　　**7.** $-4\sqrt{3}$　　**9.** $-2\sqrt{2a}$　　**11.** $19\sqrt{7}$　　**13.** $-4\sqrt{5}$

15. $23\sqrt{3}-6\sqrt{2}$　　**17.** $\dfrac{7}{3}\sqrt{15}$　　**19.** 0　　**21.** $13\sqrt[3]{3}$　　**23.** $\sqrt[4]{2}$　　**25.** $(a-2b^2)\sqrt{ab}$　　**27.** $(3-2a)\sqrt{10}$

29. $(2b-a)\sqrt[3]{3a^2b}$　　**31.** $\dfrac{(a^2-c^3)\sqrt{ac}}{a^2c^3}$　　**33.** $\dfrac{(a-2b)\sqrt[3]{ab^2}}{ab}$　　**35.** $\dfrac{2b\sqrt{a^2-b^2}}{b^2-a^2}$

37. $15\sqrt{3}-11\sqrt{5}=1.384\,014\,4$　　**39.** $\dfrac{1}{6}\sqrt{6}=0.408\,248\,3$　　**41.** $3\sqrt{3}$　　**43.** $18+3\sqrt{2}=22.2$ dm

Exercises 10-5, page 307

1. $\sqrt{30}$　　**3.** $2\sqrt{3}$　　**5.** 2　　**7.** $2\sqrt[3]{2}$　　**9.** 50　　**11.** 16　　**13.** $\dfrac{1}{3}\sqrt{30}$　　**15.** $\dfrac{1}{33}\sqrt{165}$　　**17.** $\sqrt{6}-\sqrt{15}$

19. $8-12\sqrt{3}$　　**21.** -1　　**23.** $39-12\sqrt{3}$　　**25.** $48+9\sqrt{15}$　　**27.** $66+13\sqrt{11x}-5x$　　**29.** $a\sqrt{b}+c\sqrt{ac}$

31. $5n\sqrt{3}+10\sqrt{mn}$　　**33.** $2a-3b+2\sqrt{2ab}$　　**35.** $\sqrt{6}-\sqrt{10}-2$　　**37.** $\sqrt[8]{72}$　　**39.** $c^{12}\sqrt{a^3b^7c^4}$　　**41.** 1

43. $2x\sqrt{2x}+x\sqrt[8]{8y^4}-2\sqrt[8]{x^3y^2}-y$　　**45.** $\dfrac{2-a-a^2}{a}$　　**47.** $4x^2+x-2y-4x\sqrt{x-2y}$ (valid for $x \geq 2y$)

49. $-1-\sqrt{66}=-9.124\,038\,4$　　**51.** $158+9\sqrt{182}=279.416\,64$　　**53.** $\dfrac{2x+1}{\sqrt{x}}$　　**55.** $\dfrac{5x^2+2x}{\sqrt{2x+1}}$

57. $(1-\sqrt{2})^2-2(1-\sqrt{2})-1=1-2\sqrt{2}+2-2+2\sqrt{2}-1=0$

59. $[\tfrac{1}{2}(\sqrt{b^2-4k^2}-b)]^2+b[\tfrac{1}{2}(\sqrt{b^2-4k^2}-b)]+k^2=\tfrac{1}{4}(b^2-4k^2)-\tfrac{b}{2}\sqrt{b^2-4k^2}+\tfrac{1}{4}b^2+\tfrac{b}{2}\sqrt{b^2-4k^2}-\tfrac{1}{2}b^2+k^2=0$

Exercises 10-6, page 310

1. $\sqrt{7}$　　**3.** $\dfrac{1}{2}\sqrt{14}$　　**5.** $\dfrac{1}{6}\sqrt[3]{9x^2}$　　**7.** $\dfrac{\sqrt[8]{200}}{2}$　　**9.** $\dfrac{1}{2}\sqrt[8]{4a^3}$　　**11.** $\dfrac{2-\sqrt{6}}{2}$　　**13.** $\dfrac{a\sqrt{?}-b\sqrt{a}}{a}$

15. $\dfrac{3\sqrt{a}-\sqrt{3b}}{3}$　　**17.** $\dfrac{1}{4}(\sqrt{7}-\sqrt{3})$　　**19.** $\dfrac{1}{23}(5\sqrt{7}+\sqrt{14})$　　**21.** $-\dfrac{3}{8}(\sqrt{5}+3)$　　**23.** $\dfrac{1}{13}(9+\sqrt{3})$

25. $\frac{1}{11}(\sqrt{7} + 3\sqrt{2} - 6 - \sqrt{14})$ **27.** $\frac{1}{13}(4 - \sqrt{3})$ **29.** $\frac{1}{17}(-56 + 9\sqrt{15})$ **31.** $\frac{1}{14}(4 - \sqrt{2})$

33. $\frac{2x + 2\sqrt{xy}}{x - y}$ **35.** $\frac{8(3\sqrt{a} + 2\sqrt{b})}{9a - 4b}$ **37.** $\frac{2c + 4d\sqrt{2c} + 3d^2}{2c - d^2}$ **39.** $-\frac{\sqrt{x^2 - y^2} + \sqrt{x^2 + xy}}{y}$

41. $14 - 2\sqrt{42} = 1.038\ 518\ 6$ **43.** $-\frac{16 + 5\sqrt{30}}{26} = -1.668\ 697\ 2$ **45.** $-\frac{1}{\sqrt{30} - 2\sqrt{3}}$ **47.** $\frac{1}{\sqrt{x + h} + \sqrt{x}}$

49. $\frac{\sqrt{3gsw}}{6w}$ **51.** $\frac{\sqrt{2g}(\sqrt{h_2} + \sqrt{h_1})}{2g(h_2 - h_1)}$

Review Exercises for Chapter 10, page 312

1. $\frac{2}{a^2}$ **3.** $\frac{2d^3}{c}$ **5.** 375 **7.** $\frac{1}{8000}$ **9.** $\frac{t^4}{9}$ **11.** -28 **13.** $64a^2b^5$ **15.** $-8m^9n^6$ **17.** $\frac{2y - x^2}{x^2 y}$

19. $\frac{2y}{x + 2y}$ **21.** $\frac{b}{ab - 3}$ **23.** $\frac{(x^3 y^3 - 1)^{1/3}}{y}$ **25.** $4a(a^2 + 4)^{1/2}$ **27.** $\frac{-2(x + 1)}{(x - 1)^3}$ **29.** $2\sqrt{17}$

31. $b^2 c\sqrt{ab}$ **33.** $3ab^2\sqrt{a}$ **35.** $2tu\sqrt{21st}$ **37.** $\frac{5\sqrt{2s}}{2s}$ **39.** $\frac{1}{9}\sqrt{33}$ **41.** $mn^2\sqrt[3]{8m^2n}$ **43.** $\sqrt{2}$

45. $14\sqrt{2}$ **47.** $-7\sqrt{7}$ **49.** $3ax\sqrt{2x}$ **51.** $(2a + b)\sqrt[3]{a}$ **53.** $10 - \sqrt{55}$ **55.** $4\sqrt{3} - 4\sqrt{5}$

57. $-45 - 7\sqrt{17}$ **59.** $42 - 7\sqrt{7a} - 3a$ **61.** $\frac{6x + \sqrt{3xy}}{12x - y}$ **63.** $-\frac{8 + \sqrt{6}}{29}$ **65.** $\frac{13 - 2\sqrt{35}}{29}$

67. $\frac{6x - 13a\sqrt{x} + 5a^2}{9x - 25a^2}$ **69.** $\sqrt{4b^2 + 1}$ **71.** $\frac{15 - 2\sqrt{15}}{4}$ **73.** $2\sqrt{13} + 5\sqrt{6} = 19.458\ 551$

75. $51 - 7\sqrt{105} = -20.728\ 655$ **77.** 6.0% **79.** (a) $v = k(P/W)^{1/3}$, (b) $v = \frac{k\sqrt[3]{PW^2}}{W}$ **81.** $\frac{n_1^2 n_2^2 v}{n_1^2 - n_2^2}$

83. $T = \frac{2\sqrt{2\pi\mu\ell I}}{\mu a^2}$ **85.** $6\sqrt{2}$ cm **87.** $\frac{\ell\sqrt{2a\ell + a^2}}{2\ell + a}$

Exercises 11-1, page 319

1. $9j$ **3.** $-2j$ **5.** $0.6j$ **7.** $2j\sqrt{2}$ **9.** $\frac{1}{2}j\sqrt{7}$ **11.** $-j\sqrt{\frac{2}{5}} = -\frac{1}{5}j\sqrt{10}$ **13.** $-7; 7$ **15.** $4; -4$
17. $-j$ **19.** 1 **21.** 0 **23.** $-2j$ **25.** $2 + 3j$ **27.** $-7j$ **29.** $6 + 2j$ **31.** $-2 + 3j$
33. $3\sqrt{2} - 2j\sqrt{2}$ **35.** -1 **37.** $6 + 7j$ **39.** $-2j$ **41.** $x = 2, y = -2$ **43.** $x = 10, y = -6$
45. $x = 0, y = -1$ **47.** $x = -2, y = 3$ **49.** Yes **51.** It is a real number.

Exercises 11-2, page 322

1. $5 - 8j$ **3.** $-9 + 6j$ **5.** $7 - 5j$ **7.** $-5j$ **9.** -1 **11.** $-8 + 21j$ **13.** $7 + 49j$
15. $-22.4 + 6.4j$ **17.** $22 + 3j$ **19.** $-42 - 6j$ **21.** $-18j\sqrt{2}$ **23.** $25j\sqrt{5}$ **25.** $3j\sqrt{3}$
27. $-4\sqrt{3} + 6j\sqrt{7}$ **29.** $-28j$ **31.** $3\sqrt{7} + 3j$ **33.** $-40 - 42j$ **35.** $-2 - 2j$ **37.** $\frac{1}{29}(-30 + 12j)$
39. $0.075 + 0.025j$ **41.** $-\frac{1}{3}(1 + j)$ **43.** $\frac{1}{11}(-13 + 8j\sqrt{2})$ **45.** $\frac{1}{3}(2 + 5j\sqrt{2})$ **47.** $\frac{1}{5}(-1 + 3j)$
49. $(-1 + j)^2 + 2(-1 + j) + 2 = 1 - 2j + j^2 - 2 + 2j + 2 = 0$ **51.** 13 **53.** $-\frac{1}{13}(5 + 12j)$
55. $281 + 35.2j$ volts **57.** $(a + bj) + (a - bj) = 2a$ **59.** $(a + bj) - (a - bj) = 2bj$

Exercises 11-3, page 324

1.

3.

5. $5 + 4j$

7. $8 + j$

84

9. $4 + 4j$

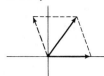

11. $-3j$

13. $-1 + 4j$

15. $-2 + 3j$

17. $4.5 + 2.0j$

19. $4 - 11j$

21. $-2j$

23. $-13 + j$

25.

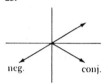

27.

29.

31.

Exercises 11-4, page 328

1. $10(\cos 36.9° + j \sin 36.9°)$

3. $5(\cos 306.9° + j \sin 306.9°)$

5. $3.61(\cos 123.7° + j \sin 123.7°)$

7. $6.00(\cos 203.6° + j \sin 203.6°)$

9. $2(\cos 60° + j \sin 60°)$

11. $8.062(\cos 295.84° + j \sin 295.84°)$

13. $3(\cos 180° + j \sin 180°)$

15. $9(\cos 90° + j \sin 90°)$

17. $2.94 + 4.05j$

19. $-1.39 + 0.800j$

21. -6

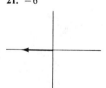

23. 8

25. $11.97 - 3.082j$

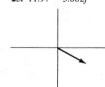

27. $-0.500 - 0.866j$

29. $-4.71 + 0.595j$ **31.** $-0.6052 - 0.7096j$ **33.** $7.32j$ **35.** $-6.961 + 86.14j$

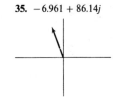

37. $R = 42.7$ N, $\theta = 306.8°$ **39.** $2.51 + 12.1j$ V/m

Exercises 11-5, page 332

1. $3.00e^{1.05j}$ **3.** $4.50e^{4.93j}$ **5.** $375.5e^{1.666j}$ **7.** $0.515e^{3.46j}$ **9.** $4.06e^{-1.07j} = 4.06e^{5.21j}$ **11.** $9245e^{5.172j}$
13. $5.00e^{5.36j}$ **15.** $3.61e^{2.55j}$ **17.** $6.37e^{0.386j}$ **19.** $825.7e^{3.836j}$ **21.** $3.00\underline{/28.6°}$; $2.63 + 1.44j$
23. $4.64\underline{/106.0°}$; $-1.28 + 4.46j$ **25.** $3.20\underline{/310.0°}$; $2.06 - 2.45j$ **27.** $0.1724\underline{/136.99°}$; $-0.1261 + 0.1176j$
29. $391e^{0.285j}$ ohms; 391 Ω **31.** $2.11 - 5.43j$ cm

Exercises 11-6, page 338

1. $8(\cos 80° + j \sin 80°)$ **3.** $3(\cos 250° + j \sin 250°)$ **5.** $2(\cos 35° + j \sin 35°)$ **7.** $2.4(\cos 110° + j \sin 110°)$
9. $8(\cos 105° + j \sin 105°)$ **11.** $256(\cos 0° + j \sin 0°)$ **13.** $0.305 + 1.70j = 1.73\underline{/79.8°}$
15. $-5134 + 10,570j = 11,750\ (\cos 115.91° + j \sin 115.91°)$ **17.** $65.0(\cos 345.7° + j \sin 345.7°) = 63 - 16j$
19. $61.4(\cos 343.9° + j \sin 343.9°)$; $59 - 17j$ **21.** $2.21(\cos 71.6° + j \sin 71.6°)$; $\frac{7}{10} + \frac{21}{10}j$
23. $0.385(\cos 120.5° + j \sin 120.5°) = \frac{1}{169}(-33 + 56j)$ **25.** $625(\cos 212.5° + j \sin 212.5°) = -527 - 336j$
27. $609(\cos 281.5° + j \sin 281.5°)$; $122 - 597j$ **29.** $2(\cos 30° + j \sin 30°)$, $2(\cos 210° + j \sin 210°)$
31. $-0.364 + 1.67j$, $-1.26 - 1.15j$, $1.63 - 0.520j$ **33.** $1, -1, j, -j$ **35.** $3j, -\frac{3}{2}(\sqrt{3} + j), \frac{3}{2}(\sqrt{3} - j)$
37. $[\frac{1}{2}(1 - j\sqrt{3})]^3 = \frac{1}{8}[1 - 3(j\sqrt{3}) + 3(j\sqrt{3})^2 - (j\sqrt{3})^3] = \frac{1}{8}[1 - 3j\sqrt{3} - 9 + 3j\sqrt{3}] = \frac{1}{8}(-8) = -1$
39. $p = 47.9\underline{/40.5°}$ watts

Exercises 11-7, page 344

1. 12.9 V **3.** (a) 2850 Ω, (b) $37.9°$, (c) 16.4 V **5.** (a) 14.6 Ω, (b) $-90.0°$ **7.** (a) 47.8 Ω, (b) $19.8°$
9. 38.0 V **11.** 54.5 Ω, $-62.3°$ **13.** 0.682 H **15.** 376 Hz **17.** 1.30×10^{-11} F $= 13.0$ pF
19. 1.02 mW

Review Exercises for Chapter 11, page 346

1. $10 - j$ **3.** $6 + 2j$ **5.** $9 + 2j$ **7.** $-12 + 66j$ **9.** $\frac{1}{85}(21 + 18j)$ **11.** $-2 - 3j$ **13.** $\frac{1}{10}(-12 + 9j)$
15. $\frac{1}{5}(13 + 11j)$ **17.** $x = -\frac{2}{3}, y = -2$ **19.** $x = -\frac{1}{2}, y = \frac{1}{2}$
21. $3 + 11j$ **23.** $4 + 8j$ **25.** $1.41(\cos 315° + j \sin 315°) = 1.41e^{5.50j}$
27. $7.28(\cos 254.1° + j \sin 254.1°) = 7.28e^{4.43j}$
29. $4.67(\cos 76.8° + j \sin 76.8°)$; $4.67e^{1.34j}$
31. $10(\cos 0° + j \sin 0°)$; $10e^{0j}$ **33.** $-1.41 - 1.41j$
35. $-2.789 + 4.163j$ **37.** $0.19 - 0.59j$

39. $26.31 - 6.427j$ **41.** $1.94 + 0.495j$ **43.** $-8.346 + 23.96j$ **45.** $15(\cos 84° + j \sin 84°)$ **47.** $20\underline{/263°}$
49. $8(\cos 59° + j \sin 59°)$ **51.** $14.29\underline{/133.61°}$ **53.** $1.26\underline{/59.7°}$ **55.** $9682\underline{/249.52°}$
57. $1024(\cos 160° + j \sin 160°)$ **59.** $27\underline{/331.5°}$ **61.** $32(\cos 270° + j \sin 270°) = -32j$
63. $\frac{625}{2}(\cos 270° + j \sin 270°) = -\frac{625}{2}j$ **65.** $1.00 + 1.73j, -2, 1.00 - 1.73j$
67. $\cos 67.5° + j \sin 67.5°, \cos 157.5° + j \sin 157.5°, \cos 247.5° + j \sin 247.5°, \cos 337.5° + j \sin 337.5°$ **69.** 60 V

71. $-21.6°$ **73.** 22.9 Hz **75.** 814 N, $317.5°$ **77.** $\dfrac{u - j\omega n}{u^2 + \omega^2 n^2}$ **79.** $e^{j\pi} = \cos \pi + j \sin \pi = -1$

Exercises 12-1, page 352

1. 3 **3.** $\frac{1}{81}$ **5.** $\log_3 27 = 3$ **7.** $\log_4 256 = 4$ **9.** $\log_4 (\frac{1}{16}) = -2$ **11.** $\log_2 (\frac{1}{64}) = -6$
13. $\log_8 2 = \frac{1}{3}$ **15.** $\log_{1/4} (\frac{1}{16}) = 2$ **17.** $81 = 3^4$ **19.** $9 = 9^1$ **21.** $5 = 25^{1/2}$ **23.** $3 = 243^{1/5}$
25. $0.1 = 10^{-1}$ **27.** $16 = (0.5)^{-4}$ **29.** 2 **31.** -2 **33.** 343 **35.** $\frac{1}{4}$ **37.** 9 **39.** $\frac{1}{64}$ **41.** 0.2
43. -3 **45.** 3 **47.** $-\frac{1}{2}$ **49.** $t = \log_{1.1} (V/A)$ **51.** $I = I_0 10^{-D}$ **53.** $N = N_0 e^{-kt}$
55. $y = \log_3 x \to x = 3^y \to y = 3^x$

Exercises 12-2, page 355

1.

3.

5.

7.

9.

11.

13.

15.

17.

19.

21.

23.

25.

27.

29.

31. Same graph as in Example C

Exercises 12-3, page 360

1. $\log_5 x + \log_5 y$ **3.** $\log_7 5 - \log_7 a$ **5.** $3 \log_2 a$ **7.** $\log_6 a + \log_6 b + \log_6 c$ **9.** $\frac{1}{4} \log_5 y$
11. $\frac{1}{2} \log_2 x - 2 \log_2 a$ **13.** $\log_b ac$ **15.** $\log_5 3$ **17.** $\log_b x^{3/2}$ **19.** $\log_e 4n^3$ **21.** -5 **23.** 2.5
25. $\frac{1}{2}$ **27.** $\frac{3}{4}$ **29.** $2 + \log_3 2$ **31.** $-1 - \log_2 3$ **33.** $\frac{1}{2}(1 + \log_3 2)$ **35.** $4 + 3 \log_2 3$

37. $3 + \log_{10} 3$ **39.** $-2 + 3 \log_{10} 3$ **41.** $y = 2x$ **43.** $y = \dfrac{3x}{5}$ **45.** $y = \dfrac{49}{x^3}$ **47.** $y = 2(2ax)^{1/5}$

49. $y = \dfrac{2}{x}$ **51.** $y = \dfrac{x^2}{25}$ **53.** 0.602   **55.** -0.301 **57.**   **59.** $T = 65e^{-0.41t}$

Exercises 12-4, page 364

1. 2.754   **3.** -1.194 **5.** 6.966   **7.** -3.9311 **9.** 1.8649   **11.** $-0.729\,48$ **13.** -0.0104 **15.** 1.219
17. 27 400   **19.** 0.049 60   **21.** 2000.4   **23.** 0.7234   **25.** 1.4284   **27.** 0.005 788 2   **29.** 85.5
31. 94 600   **33.** 1.44×10^{341} **35.** 7.37×10^{101} **37.** 9.0607   **39.** 5.176   **41.** 15.2 dB   **43.** 1260 g

Exercises 12-5, page 368

1. 3.258   **3.** 0.4460   **5.** -0.6898 **7.** $-4.916\,23$ **9.** 1.92   **11.** 3.418   **13.** 1.933   **15.** 1.795
17. 3.940   **19.** 0.3322   **21.** $-0.008\,335$ **23.** $-4.347\,66$ **25.** 3.8104   **27.** $-0.377\,93$ **29.** 8.94
31. 1.0085   **33.** 0.4757   **35.** 6.20×10^{-11} **37.** $y = 3x$ **39.** 8.155%   **41.** 0.384 s   **43.** 21.7 s

Exercises 12-6, page 372

1. 4   **3.** -0.748 **5.** 0.587   **7.** 0.6439   **9.** 0.285   **11.** 0.203   **13.** 4.11   **15.** 14.2   **17.** $\frac{1}{4}$
19. 1.649   **21.** 3   **23.** -0.162 **25.** 5   **27.** 0.906   **29.** 4   **31.** 2   **33.** 4   **35.** 1.42   **37.** 28.0
39. 6.84 min   **41.** 3.922×10^{-4} **43.** $10^{8.25} = 1.78 \times 10^8$ **45.** $n = 20e^{-0.04t}$ **47.** $P = P_0(0.999)^t$
49. 3.35   **51.** ± 3.7 m

Exercises 12-7, page 376

1.

3.

5.

7.

9.

11.

13.

15.

17.

19.

21.

23.

25. (a) (b)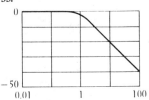

27.

29.

31.

33.

35.

Review Exercises for Chapter 12, page 379

1. 10 000   **3.** $\frac{1}{5}$ **5.** 6   **7.** $\frac{5}{3}$ **9.** 6   **11.** 100   **13.** $\log_3 2 + \log_3 x$ **15.** $2 \log_3 t$ **17.** $2 + \log_2 7$

19. $2 - \log_3 x$ **21.** $1 + \frac{1}{2} \log_4 3$ **23.** $3 + 4 \log_{10} x$ **25.** $y = \frac{4}{x}$ **27.** $y = \frac{8}{x}$ **29.** $y = \frac{15}{x}$ **31.** $y = 2x$

33.

35.

37.

39.

41. 9.42   **43.** 3.88×10^{445} **45.** 2.18   **47.** 0.728   **49.** 0.805   **51.** 4.30   **53.** 2   **55.** 30

57.

59.

61. 4   **63.** 12   **65.** $t = 12 \ln (V/1000)$

67. $t = -0.2 \ln \left(\dfrac{i}{i_0} \right)$

69.

71. $\sin\theta = \dfrac{\ell\omega^2}{3g}$ **73.** $R = 2^{C/B} - 1$ **79.**

75. $4.68\ \text{g}\cdot\text{mol/L}$ **77.** $1.13\ \text{min}$

Exercises 13-1, page 386

1. $x = 1.8,\ y = 3.6;\ x = -1.8,\ y = -3.6$ **3.** $x = 0.0,\ y = -2.0;\ x = 2.7,\ y = -0.7$ **5.** $x = 1.5,\ y = 0.2$
7. $x = 1.6,\ y = 2.5$ **9.** $x = 1.1,\ y = 2.8;\ x = -1.1,\ y = 2.8;\ x = 2.4,\ y = -1.8;\ x = -2.4,\ y = -1.8$
11. No solution   **13.** $x = -2.8,\ y = -1.0;\ x = 2.8,\ y = 1.0;\ x = 2.8,\ y = -1.0;\ x = -2.8,\ y = 1.0$
15. $x = 0.7,\ y = 0.7;\ x = -0.7,\ y = -0.7$ **17.** $x = 0.0,\ y = 0.0;\ x = 0.9,\ y = 0.8$
19. $x = -1.1,\ y = 3.0;\ x = 1.8,\ y = 0.2$ **21.** $x = 1.0,\ y = 0.0$ **23.** $x = 3.6,\ y = 1.0$
25. 4.9 km N, 1.6 km E   **27.** 2.2 A, 0.9 A

Exercises 13-2, page 391

1. $x = 0,\ y = 1;\ x = 1,\ y = 2$ **3.** $x = -\frac{19}{5},\ y = \frac{17}{5};\ x = 5,\ y = -1$ **5.** $x = 1,\ y = 0$
7. $x = \frac{2}{7}(3 + \sqrt{2}),\ y = \frac{2}{7}(-1 + 2\sqrt{2});\ x = \frac{2}{7}(3 - \sqrt{2}),\ y = \frac{2}{7}(-1 - 2\sqrt{2})$ **9.** $x = 1,\ y = 1$
11. $x = \frac{2}{3},\ y = \frac{9}{2};\ x = -3,\ y = -1$ **13.** $x = -2,\ y = 4;\ x = 2,\ y = 4$ **15.** $x = 1,\ y = 2;\ x = -1,\ y = 2$
17. $x = 1,\ y = 0;\ x = -1,\ y = 0;\ x = \frac{1}{2}\sqrt{6},\ y = \frac{1}{2};\ x = -\frac{1}{2}\sqrt{6},\ y = \frac{1}{2}$
19. $x = \sqrt{19},\ y = \sqrt{6};\ x = \sqrt{19},\ y = -\sqrt{6};\ x = -\sqrt{19},\ y = \sqrt{6};\ x = -\sqrt{19},\ y = -\sqrt{6}$
21. $x = \frac{1}{11}\sqrt{22},\ y = \frac{1}{11}\sqrt{770};\ x = \frac{1}{11}\sqrt{22},\ y = -\frac{1}{11}\sqrt{770};\ x = -\frac{1}{11}\sqrt{22},\ y = \frac{1}{11}\sqrt{770};\ x = -\frac{1}{11}\sqrt{22},\ y = -\frac{1}{11}\sqrt{770}$
23. $x = -5,\ y = -2;\ x = -5,\ y = 2;\ x = 5,\ y = -2;\ x = 5,\ y = 2$ **25.** $v_1 = -4.0\ \text{m/s},\ v_2 = 2.5\ \text{m/s}$
27. 1.5 cm, 1.4 cm   **29.** 11.8 m, 38.2 m   **31.** 80 km/h

Exercises 13-3, page 395

1. $-3, -2, 2, 3$ **3.** $\frac{3}{2}, -\frac{3}{2}, j, -j$ **5.** $-\frac{1}{2}, \frac{1}{4}$ **7.** $-\frac{1}{2}, \frac{1}{2}, \frac{1}{6}j\sqrt{6}, -\frac{1}{6}j\sqrt{6}$ **9.** $1, \frac{25}{4}$ **11.** $\frac{64}{729}, 1$
13. $-27, 125$ **15.** 256 **17.** 5 **19.** $-2, -1, 3, 4$ **21.** 18
23. $1, -2, 1 + j\sqrt{3}, 1 - j\sqrt{3}, \frac{1}{2}(-1 + j\sqrt{3}), -\frac{1}{2}(1 + j\sqrt{3})$ **25.** $R_1 = 2.62\ \Omega,\ R_2 = 1.62\ \Omega$ **27.** 16 m by 12 m

Exercises 13-4, page 398

1. 12 **3.** 2 **5.** $\frac{2}{3}$ **7.** $\frac{1}{2}(7 + \sqrt{5})$ **9.** -1 **11.** 32 **13.** 16 **15.** 9 **17.** 12 **19.** 16 **21.** $\frac{1}{2}$
23. $7, -1$ **25.** 0 **27.** 5 **29.** 6 **31.** 258 **33.** $L = \dfrac{1}{4\pi^2 f^2 C}$ **35.** $\dfrac{k^2}{2n(1 - k)}$ **37.** 1.5 m, 2.0 m, 2.5 m
39. 3.4 km

Review Exercises for Chapter 13, page 399

1. $x = -0.9,\ y = 3.5;\ x = 0.8,\ y = 2.6$ **3.** $x = 2.0,\ y = 0.0;\ x = 1.6,\ y = 0.6$
5. $x = 0.8,\ y = 1.6;\ x = -0.8,\ y = 1.6$ **7.** $x = -1.2,\ y = 2.7;\ x = 1.2,\ y = 2.7$
9. $x = 0.0,\ y = 0.0;\ x = 2.4,\ y = 0.9$ **11.** $x = 0,\ y = 0;\ x = 2,\ y = 16$ **13.** $x = \sqrt{2},\ y = 1;\ x = -\sqrt{2},\ y = 1$
15. $x = \frac{1}{12}(1 + \sqrt{97}),\ y = \frac{1}{18}(5 - \sqrt{97});\ x = \frac{1}{12}(1 - \sqrt{97}),\ y = \frac{1}{18}(5 + \sqrt{97})$
17. $x = 7,\ y = 5;\ x = 7,\ y = -5;\ x = -7,\ y = 5;\ x = -7,\ y = -5$ **19.** $x = -2,\ y = 2;\ x = \frac{2}{3},\ y = \frac{10}{9}$
21. $-4, -2, 2, 4$ **23.** $1, 16$ **25.** $\frac{1}{3}, -\frac{1}{7}$ **27.** $\frac{25}{4}$ **29.** $-\frac{1}{2}, -2$ **31.** 11 **33.** $1, 4$ **35.** 8 **37.** $\frac{9}{16}$
39. $\frac{1}{3}(11 - 4\sqrt{15})$ **41.** $\ell = \frac{1}{2}(-1 + \sqrt{1 + 16\pi^2 L^2/h^2})$ **43.** $m = \frac{1}{2}(-y \pm \sqrt{2s^2 - y^2})$ **45.** 0.40 s, 0.80 s
47. 70 m   **49.** $Z = 1.09\ \Omega,\ X = 0.738\ \Omega$ **51.** 37.7 cm, 29.7 cm   **53.** 3.57 cm   **55.** 26.3 km/h, 32.3 km/h

Exercises 14-1, page 403

1. 0 **3.** 0 **5.** -40 **7.** -4 **9.** 8 **11.** 183 **13.** -28 **15.** 14 **17.** Yes   **19.** Yes
21. No   **23.** No   **25.** Yes   **27.** Yes   **29.** $(4x^3 + 8x^2 - x - 2) \div (2x - 1) = 2x^2 + 5x + 2$; No   **31.** -7

Exercises 14-2, page 408

1. $x^2 + 3x + 2,\ R = 0$ **3.** $x^2 - x + 3,\ R = 0$ **5.** $2x^4 - 4x^3 + 8x^2 - 17x + 42,\ R = -40$
7. $3x^3 - x + 2,\ R = -4$ **9.** $x^2 + x - 4,\ R = 8$ **11.** $x^3 - 3x^2 + 10x - 45,\ R = 183$

13. $2x^3 - x^2 - 4x - 12, R = -28$ **15.** $x^4 + 2x^3 + x^2 + 7x + 4, R = 14$
17. $x^5 + 2x^4 + 4x^3 + 8x^2 + 18x + 36, R = 66$ **19.** $x^6 + 2x^5 + 4x^4 + 8x^3 + 16x^2 + 32x + 64, R = 0$
21. Yes   **23.** No   **25.** No   **27.** No   **29.** Yes   **31.** No   **33.** Yes   **35.** Yes

Exercises 14-3, page 413

(Note: Unknown roots listed)
1. $-2, -1$ **3.** $-2, 3$ **5.** $-2, -2$ **7.** $-j, -\frac{2}{3}$ **9.** $2j, -2j$ **11.** $-1, -3$ **13.** $-2, 1$
15. $1 - j, \frac{1}{2}, -2$ **17.** $\frac{1}{4}(1 + \sqrt{17}), \frac{1}{4}(1 - \sqrt{17})$ **19.** $2, -3$ **21.** $-j, -1, 1$ **23.** $-2j, 3, -3$

Exercises 14-4, page 419

1. $1, -1, -2$ **3.** $2, -1, -3$ **5.** $\frac{1}{2}, 5, -3$ **7.** $\frac{1}{3}, -3, -1$ **9.** $-2, -2, 2 \pm \sqrt{3}$ **11.** $2, 4, -1, -3$
13. $1, -\frac{1}{2}, 1 \pm \sqrt{3}$ **15.** $\frac{1}{2}, -\frac{2}{3}, -3, -\frac{1}{2}$ **17.** $2, 2, -1, -1, -3$ **19.** $\frac{1}{2}, 1, 1, j, -j$ **21.** 0.59
23. -0.77 **25.** 2 cm^3 **27.** $0, L$ **29.** 1.5 kg, 4.1 kg   **31.** $2 \Omega, 3 \Omega, 6 \Omega$ **33.** 3.0 cm, 4.0 cm
35. $-1.86, 0.68, 3.18$

Review Exercises for Chapter 14, page 420

1. 1   **3.** -107 **5.** Yes   **7.** No   **9.** $x^2 + 4x + 10, R = 11$ **11.** $2x^2 - 7x + 10, R = -17$
13. $x^3 - 3x^2 - 4, R = -4$ **15.** $2x^4 + 10x^3 + 4x^2 + 21x + 105, R = 516$ **17.** No   **19.** Yes
21. (Unlisted roots) $\frac{1}{2}(-5 \pm \sqrt{17})$ **23.** (Unlisted roots) $\frac{1}{3}(-1 \pm j\sqrt{14})$ **25.** (Unlisted roots) $4, -3$
27. (Unlisted roots) $-j, \frac{1}{2}, -\frac{3}{2}$ **29.** (Unlisted roots) $2, -2$
31. (Unlisted roots) $-2 - j, \frac{1}{2}(-1 + j\sqrt{3}), -\frac{1}{2}(1 + j\sqrt{3})$ **33.** $1, 2, -4$ **35.** $-1, -1, \frac{5}{2}$ **37.** $\frac{5}{3}, -\frac{1}{2}, -1$
39. $\frac{1}{2}, -1, j\sqrt{2}, -j\sqrt{2}$ **41.** 4   **43.** $1, 0.4, 1.5, -0.6$ (last three are irrational)   **45.** 1.91   **47.** April
49. 0.75 cm   **51.** 2 cm   **53.** 2 cm, 3 cm, 4 cm   **55.** 8 m, 15 m

Exercises 15-1, page 428

1. 39   **3.** 30   **5.** 50   **7.** $-47\,416$ **9.** -6 **11.** -86 **13.** 118   **15.** -2
17. $x = 2, y = -1, z = 3$ **19.** $x = -1, y = \frac{1}{3}, z = -\frac{1}{2}$ **21.** $x = -1, y = 0, z = 2, t = 1$
23. $x = 1, y = 2, z = -1, t = 3$ **25.** $\frac{2}{7}\text{A}, \frac{18}{7}\text{A}, -\frac{8}{7}\text{A}, -\frac{12}{7}\text{A}$ **27.** 500, 300, 700

Exercises 15-2, page 434

1. -60 **3.** -56 **5.** 0   **7.** 0   **9.** 57   **11.** -124 **13.** -13 **15.** -118 **17.** -72 **19.** 0
21. $x = 0, y = -1, z = 4$ **23.** $x = \frac{1}{3}, y = -\frac{1}{2}, z = 1$ **25.** $x = 2, y = -1, z = -1, t = 3$
27. $x = 1, y = 2, z = -1, t = -2$ **29.** $\frac{33}{16}\text{A}, \frac{11}{8}\text{A}, -\frac{5}{8}\text{A}, -\frac{15}{8}\text{A}, -\frac{15}{16}\text{A}$
31. ppm of SO_2, NO, NO_2, CO: 0.5, 0.3, 0.2, 5.0

Exercises 15-3, page 439

1. $a = 1, b = -3, c = 4, d = 7$ **3.** $x = 2, y = 3$ **5.** Elements cannot be equated; different number of rows

7. $x = 4, y = 6, z = 9$ **9.** $\begin{pmatrix} 1 & 10 \\ 0 & 2 \end{pmatrix}$ **11.** $\begin{pmatrix} -5 & 0 \\ 11 & 71 \\ 11 & -5 \end{pmatrix}$ **13.** $\begin{pmatrix} 0 & 9 & -13 & 3 \\ 6 & -7 & 7 & 0 \end{pmatrix}$

15. Cannot be added   **17.** $\begin{pmatrix} -1 & 13 & -20 & 3 \\ 8 & -13 & 6 & 2 \end{pmatrix}$ **19.** $\begin{pmatrix} -3 & -6 & 5 & -6 \\ -6 & -4 & -17 & 6 \end{pmatrix}$

21. $A + B = B + A = \begin{pmatrix} 3 & 1 & 0 & 7 \\ 5 & -3 & -2 & 5 \\ 10 & 10 & 8 & 0 \end{pmatrix}$ **23.** $-(A - B) = B - A = \begin{pmatrix} 5 & -3 & -6 & -7 \\ 5 & 3 & 0 & -3 \\ -8 & 12 & 8 & 4 \end{pmatrix}$

25. $v_1 = 58.2$ m/s, $v_2 = 46.3$ m/s   **27.** $\begin{pmatrix} 24 & 18 & 0 & 0 \\ 15 & 12 & 9 & 0 \\ 0 & 9 & 15 & 18 \end{pmatrix}$

Exercises 15-4, page 444

1. $(-8 \quad -12)$ **3.** $\begin{pmatrix} -15 & 15 & -26 \\ 8 & 5 & -13 \end{pmatrix}$ **5.** $\begin{pmatrix} 29 \\ -29 \end{pmatrix}$ **7.** $\begin{pmatrix} -13.21 & -9.52 \\ 11.50 & 0.00 \\ 6.89 & 0.68 \end{pmatrix}$

9. $\begin{pmatrix} 33 & -22 & 7 \\ 31 & -12 & 5 \\ 15 & 13 & -1 \\ 50 & -41 & 12 \end{pmatrix}$ **11.** $\begin{pmatrix} -62 & 68 \\ 73 & -27 \end{pmatrix}$ **13.** $AB = (40)$, $BA = \begin{pmatrix} -1 & 3 & -8 \\ 5 & -15 & 40 \\ 7 & -21 & 56 \end{pmatrix}$

15. $AB = \begin{pmatrix} -5 \\ 10 \end{pmatrix}$, BA not defined **17.** $AI = IA = A$ **19.** $AI = IA = A$ **21.** $B = A^{-1}$ **23.** $B = A^{-1}$

25. Yes **27.** No **29.** $\begin{pmatrix} -1 & 0 \\ 0 & -1 \end{pmatrix}\begin{pmatrix} -1 & 0 \\ 0 & -1 \end{pmatrix} = \begin{pmatrix} 1 & 0 \\ 0 & 1 \end{pmatrix}$ **31.** $A^2 - I = (A+I)(A-I) = \begin{pmatrix} 15 & 28 \\ 21 & 36 \end{pmatrix}$

33. $\begin{pmatrix} 0 & -j \\ j & 0 \end{pmatrix}\begin{pmatrix} 0 & -j \\ j & 0 \end{pmatrix} = \begin{pmatrix} 1 & 0 \\ 0 & 1 \end{pmatrix}$ **35.** $v_2 = v_1$, $i_2 = -v_1/R + i_1$

Exercises 15-5, page 449

1. $\begin{pmatrix} -2 & -\frac{5}{2} \\ -1 & -1 \end{pmatrix}$ **3.** $\begin{pmatrix} -\frac{1}{3} & \frac{1}{6} \\ \frac{2}{15} & \frac{1}{30} \end{pmatrix}$ **5.** $\begin{pmatrix} \frac{3}{4} & \frac{1}{2} \\ -\frac{1}{4} & 0 \end{pmatrix}$ **7.** $\begin{pmatrix} -\frac{8}{283} & -\frac{9}{566} \\ \frac{13}{1415} & \frac{5}{283} \end{pmatrix}$ **9.** $\begin{pmatrix} -3 & 2 \\ 2 & -1 \end{pmatrix}$ **11.** $\begin{pmatrix} -\frac{1}{2} & -2 \\ \frac{1}{2} & 1 \end{pmatrix}$

13. $\begin{pmatrix} \frac{2}{9} & -\frac{5}{9} \\ \frac{1}{9} & \frac{2}{9} \end{pmatrix}$ **15.** $\begin{pmatrix} \frac{3}{8} & \frac{1}{16} \\ -\frac{1}{4} & \frac{1}{8} \end{pmatrix}$ **17.** $\begin{pmatrix} -18 & -7 & 5 \\ -3 & -1 & 1 \\ -5 & -2 & 1 \end{pmatrix}$ **19.** $\begin{pmatrix} 3 & -4 & -1 \\ -4 & 5 & 2 \\ 2 & -3 & -1 \end{pmatrix}$

21. $\begin{pmatrix} 2 & 4 & \frac{7}{2} \\ -1 & -2 & -\frac{3}{2} \\ 1 & 1 & \frac{1}{2} \end{pmatrix}$ **23.** $\begin{pmatrix} \frac{5}{2} & -2 & -2 \\ -1 & 1 & 1 \\ \frac{7}{4} & -\frac{3}{2} & -1 \end{pmatrix}$ **25.** $\begin{pmatrix} 2 & 4 & \frac{7}{2} \\ -1 & -2 & -\frac{3}{2} \\ 1 & 1 & \frac{1}{2} \end{pmatrix}$

27. $\begin{pmatrix} \frac{5}{2} & -2 & -2 \\ -1 & 1 & 1 \\ \frac{7}{4} & -\frac{3}{2} & -1 \end{pmatrix}$ **29.** $\frac{1}{ad-bc}\begin{pmatrix} ad - bc & -ba + ab \\ cd - dc & -bc + ad \end{pmatrix} = \begin{pmatrix} 1 & 0 \\ 0 & 1 \end{pmatrix}$ **31.** $v_1 = a_{22}i_1 - a_{12}i_2$ $v_2 = -a_{21}i_1 + a_{11}i_2$

Exercises 15-6, page 453

1. $x = \frac{1}{2}$, $y = 3$ **3.** $x = \frac{1}{2}$, $y = -\frac{5}{2}$ **5.** $x = -4$, $y = 2$, $z = -1$ **7.** $x = -1$, $y = 0$, $z = 3$
9. $x = 1$, $y = 2$ **11.** $x = -\frac{3}{2}$, $y = -2$ **13.** $x = -3$, $y = -\frac{1}{2}$ **15.** $x = 1.6$, $y = -2.5$
17. $x = 2$, $y = -4$, $z = 1$ **19.** $x = 2$, $y = -\frac{1}{2}$, $z = 3$ **21.** $A = 118$ N, $B = 186$ N **23.** 6.4 L, 1.6 L, 2.0 L

Review Exercises for Chapter 15, page 455

1. 6 **3.** 186 **5.** -438 **7.** 44 **9.** 6 **11.** 186 **13.** -438 **15.** 44 **17.** -9 **19.** -44

21. $a = 4$, $b = -1$ **23.** $x = 2$, $y = -3$, $z = \frac{5}{2}$, $a = -1$, $b = -\frac{7}{2}$, $c = \frac{1}{2}$ **25.** $\begin{pmatrix} 1 & -3 \\ 8 & -5 \\ -8 & -2 \\ 3 & -10 \end{pmatrix}$ **27.** $\begin{pmatrix} 3 & 0 \\ -12 & 18 \\ 9 & 6 \\ -3 & 21 \end{pmatrix}$

29. Cannot be subtracted **31.** $\begin{pmatrix} 7 & -6 \\ -4 & 20 \\ -1 & 6 \\ 1 & 15 \end{pmatrix}$ **33.** $\begin{pmatrix} 13 \\ -13 \end{pmatrix}$ **35.** $\begin{pmatrix} 34 & 11 & -5 \\ 2 & -8 & 10 \\ -1 & -17 & 20 \end{pmatrix}$ **37.** $\begin{pmatrix} -2 & \frac{5}{2} \\ -1 & 1 \end{pmatrix}$

39. $\begin{pmatrix} \frac{2}{15} & \frac{1}{60} \\ -\frac{1}{15} & \frac{7}{60} \end{pmatrix}$ **41.** $\begin{pmatrix} 11 & 10 & 3 \\ -4 & -4 & -1 \\ 3 & 3 & 1 \end{pmatrix}$ **43.** $\begin{pmatrix} \frac{1}{2} & -\frac{1}{2} & -1 \\ -3 & 2 & 1 \\ -4 & 3 & 2 \end{pmatrix}$ **45.** $x = -3$, $y = 1$ **47.** $x = 10$, $y = -15$

49. $x = -1$, $y = -3$, $z = 0$ **51.** $x = 1$, $y = \frac{1}{2}$, $z = -\frac{1}{3}$ **53.** $x = 3$, $y = 1$, $z = -1$

55. $x = 1$, $y = 2$, $z = -3$, $t = 1$ **57.** $2\sqrt{2}$ **59.** 0 **61.** $N^{-1} = -N = \begin{pmatrix} 0 & 1 \\ -1 & 0 \end{pmatrix}$

63. $(A+B)(A-B) = \begin{pmatrix} -6 & 2 \\ 4 & 2 \end{pmatrix}$, $A^2 - B^2 = \begin{pmatrix} -10 & -4 \\ 8 & 6 \end{pmatrix}$ **65.** $\begin{pmatrix} \frac{1}{2} & \frac{1}{3} \\ 0 & \frac{1}{6} \end{pmatrix} = \frac{1}{2}\begin{pmatrix} 1 & \frac{2}{3} \\ 0 & \frac{1}{3} \end{pmatrix}$ **67.** $R_1 = 4\,\Omega$, $R_2 = 6\,\Omega$

69. $F = 303$ N, $T = 175$ N   **71.** 0.20 h after police pass intersection   **73.** 30 g, 50 g, 20 g

75. 2.0 h, 1.5 h, 1.0 h, 1.0 h   **77.** $\begin{pmatrix} 15\,000 & 10\,000 \\ 20\,000 & 18\,000 \\ 8\,000 & 30\,000 \end{pmatrix} + \begin{pmatrix} 18\,000 & 12\,000 \\ 30\,000 & 22\,000 \\ 12\,000 & 40\,000 \end{pmatrix} = \begin{pmatrix} 33\,000 & 22\,000 \\ 50\,000 & 40\,000 \\ 20\,000 & 70\,000 \end{pmatrix}$

79. $(R_1 + R_2)i_1 - R_2 i_2 = 6$
   $- R_2 i_1 + (R_1 + R_2)i_2 = 0$

Exercises 16-1, page 465

1. $7 < 12$ **3.** $20 < 45$ **5.** $-4 > -9$ **7.** $16 < 81$ **9.** $x > -2$ **11.** $x \le 4$ **13.** $1 < x < 7$
15. $x < -9$ or $x \ge -4$ **17.** $x < 1$ or $3 < x \le 5$ **19.** $-2 < x < 2$ or $3 \le x < 4$
21. x is greater than 0 and less than or equal to 2
23. x is less than -1, or greater than or equal to 1 and less than 2

25. ○—— 3 **27.** ●—○ 1 3 **29.** ●—○ 0 5 **31.** ○—○ ●—○ -1 1 4

33. ○—○—○—● -3 -1 1 3 **35.** ——○—— -3 **37.** $d > 5 \times 10^{12}$ km   ├——○——→ 0 5×10^{12} **39.** $29\,000 < v < 40\,000$ km/h   ├——○——○— 0 $29\,000$ $40\,000$

41. $0 < n \le 2565$ steps   **43.** $E = 0$ for $0 \le r < a$
   $E = k/r^2$ for $r \ge a$

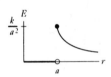

Exercises 16-2, page 468

1. $x > -1$ ○—— -1 **3.** $x < 6$ ——○ 6 **5.** $x \le -2$ ●—— -2 **7.** $x < 2$ ——○ 2 **9.** $x \le \frac{5}{2}$ ——● $\frac{5}{2}$

11. $x < -1$ ——○ -1 **13.** $x \le \frac{13}{2}$ ——● $\frac{13}{2}$ **15.** $x > -\frac{7}{9}$ ○—— $-\frac{7}{9}$ **17.** $-1 < x < 1$ ○——○ -1 1 **19.** $2 < x \le 5$ ○——● 2 5

21. $-3 \le x < -1$ ●——○ -3 -1 **23.** No values   ——┼—— 0 **25.** $0 \le t \le 6$ years   ●——● 0 6 **27.** $0 \le t < 5.0$ s   ●——○ 0 5.0

29. $0 \le x \le 500$
   $200 \le y \le 700$
   ●——● 0 200 700 **31.** $24 \le x \le 40$ L   ●——● 0 24 40

Exercises 16-3, page 474

1. $-1 < x < 1$ ○——○ -1 1 **3.** $0 \le x \le 2$ ●——● 0 2 **5.** $x \le -2, x \ge \frac{1}{3}$ ●——● -2 $\frac{1}{3}$ **7.** $\frac{1}{3} < x < \frac{1}{2}$ ○——○ $\frac{1}{3}$ $\frac{1}{2}$ **9.** $x = -2$ ● -2

11. All x ——┼—— 0 **13.** $-2 < x < 0, x > 1$ ○——○—○ -2 0 1 **15.** $-2 \le x \le -1, x \ge 1$ ●——●—● -2 -1 1 **17.** $x < 3, x > 8$ ——○ ○—— 3 8

19. $-6 < x \le \frac{3}{2}$ ○——● -6 $\frac{3}{2}$ **21.** $-1 < x < 2$ ○——○ -1 2 **23.** $-5 < x < -1, x > 7$ ○——○ ○—— -5 -1 7 **25.** $-1 < x < \frac{3}{4}, x \ge 6$ ○——○ ●— -1 $\frac{3}{4}$ 6

27. $2 < x < 4, 5 < x < 9$ ○——○ ○——○ 2 4 5 9 **29.** $x \le -2, x \ge 1$ **31.** $-1 \le x \le 0$

33.
$x > 1.3$

35.
$-1.4 < x < -0.4$

37.
$x > 1.6$

39.
$0 \le x < 0.1,$
$3.0 < x \le 2\pi$

41. $0.5 < i < 1$ A   **43.** $0 \le t < 0.68$ s   **45.** $3.0 \le w < 5.0$ mm   **47.** $0 \le t < \frac{1}{4}(\sqrt{46} - 2) = 1.20$ h

Exercises 16-4, page 477

1. $3 < x < 5$

3   5

3. $x < -2, x > \frac{2}{5}$
-2 $\frac{2}{5}$

5. $\frac{1}{6} \le x \le \frac{3}{2}$
$\frac{1}{6}$ $\frac{3}{2}$

7. $x < 0, x > \frac{3}{2}$
0    $\frac{3}{2}$

9. $-16 < x < 14$
-16 14

11. $-6.4 \le x \le -2.1$
-6.4 -2.1

13. $x < 0, x > 8$
0   8

15. $x \le \frac{1}{10}, x \ge \frac{7}{10}$
$\frac{1}{10}$ $\frac{7}{10}$

17. $-18 < x < 14$
-18 14

19. $x \le 8.4, x \ge 17.6$
8.4   17.6

21. $x < -3, -2 < x < 1, x > 2$

23. $-3 < x < -2, 1 < x < 2$

25. $|d - 0.2537| \le 0.0003$ cm   **27.** $d = 3.0x - 10.0; |d| < 6.0$ m for $1.3 < x < 5.3$ m

Exercises 16-5, page 482

1.   **3.**   **5.**   **7.**   **9.**

11.   **13.**   **15.**   **17.**   **19.**

21.   **23.**   **25.**   **27.**   **29.**

31.   **33.** 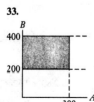   **35.**   **37.**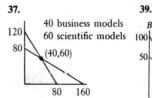

40 business models
60 scientific models
(40,60)

39.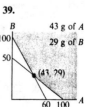

43 g of A
29 g of B
(43, 29)

Review Exercises for Chapter 16, page 484

1. $x > 6$

3. $x \le -\frac{5}{3}$

5. $x \le -\frac{29}{2}$

7. $\frac{5}{2} < x < 6$

9. $-2 < x < 1$

11. $-2 < x < \frac{1}{5}$

13. $x < -9, x > 7$

15. $-4 < x < -1, x > 1$

17. $-\frac{1}{2} < x \le 8$

19. $x < -4, \frac{1}{2} < x < 3$

21. $x = 0$

23. $x < 0, x > \frac{1}{2}$

25. $x < -1, x > 5$

27. $-2 \le x \le \frac{2}{3}$

29. $x < -\frac{4}{5}, x > 2$

31. $-6 \le x \le 14$

33.

$x < -0.68$

35.

$x < 0.7$

37. $x \le 3$

39. $x \le -4, x \ge 0$

41.

43.

45.

47.

49.

51.

53. $T \ge 4000°C$ **55.** $168 \le B \le 400$ MJ   **57.** $0.46 < i < 0.82$ A
59. $0 \le t < 0.52$ s, $2.62 < t < 3.67$ s, $5.76 < t < 6.28$ s

61.
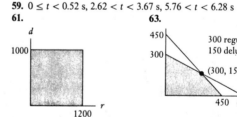

63.
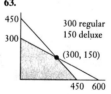
300 regular
150 deluxe
(300, 150)

Exercises 17-1, page 489

1. 6   **3.** $\frac{4}{25}$ **5.** 48   **7.** 40   **9.** 1.7   **11.** 0.41   **13.** 3.5%   **15.** 863 kg   **17.** 1.44 Ω   **19.** 0.103 m³
21. 20 000 g   **23.** 900 kJ   **25.** 286°   **27.** 12.5 m/s   **29.** 22 500 cm³   **31.** 19.67 kg
33. 100 mg, 120 mg   **35.** 3500 lines, 4900 lines

Exercises 17-2, page 495

1. $y = kz$ **3.** $s = \dfrac{k}{t^2}$ **5.** $f = k\sqrt{x}$ **7.** $w = kxy^3$ **9.** $r = \dfrac{16}{y}$ **11.** $y = \frac{1}{4}\sqrt{x}$ **13.** $s = \dfrac{7}{2\sqrt{t}}$

15. $p = \dfrac{16q}{r^3}$ **17.** 25   **19.** 50   **21.** 180   **23.** 2.56×10^5 **25.** 61 m³   **27.** $m = 0.476t$ **29.** 0.67

31. 1.4 h   **33.** 1250   **35.** 28 N   **37.** $F = 2.32Av^2$ **39.** 482 m/s   **41.** $R = \dfrac{2.60 \times 10^{-5}\ell}{A}$

43. 80.0 W   **45.** $G = \dfrac{5.9d^2}{\lambda^2}$ **47.** -6.57 cm/s²

Review Exercises for Chapter 17, page 497

1. 200   **3.** $\frac{2}{3}$ **5.** 5.6   **7.** 7.39 N/cm^2 **9.** 3.25 cm   **11.** 2.66×10^{-3} kJ   **13.** 36 000 characters

15. 140 mL   **17.** 71.9 m   **19.** 4500, 7500   **21.** $y = 3x^2$ **23.** $v = \dfrac{128x}{y^3}$ **25.** 11.7 m   **27.** 4.3 μC

29. 18.0 kW   **31.** 44.1 m   **33.** 1.4   **35.** 50.0 Hz   **37.** 2.99×10^8 m/s   **39.** 3.26 cm   **41.** 4.8 MJ
43. 5.73×10^4 m   **45.** 149%   **47.** $125.00, $600.13

Exercises 18-1, page 505

1. 4, 6, 8, 10, 12   **3.** 13, 9, 5, 1, -3 **5.** 22   **7.** -62 **9.** 37   **11.** $49b$ **13.** 440   **15.** $-\frac{85}{2}$
17. $n = 6$, $S_6 = 150$ **19.** $d = -\frac{2}{19}$, $a_{20} = -\frac{1}{3}$ **21.** $a_1 = 19$, $a_{30} = 106$ **23.** $n = 62$, $S_{62} = -4867$
25. $n = 23$, $a_{23} = 6k$ **27.** $n = 8$, $d = \frac{1}{14}(b + 2c)$ **29.** $a_1 = 36$, $d = 4$, $S_{10} = 540$
31. $a_1 = 3$, $d = -\frac{1}{3}$, $S_{10} = 15$ **33.** 5050   **35.** 100 500   **37.** 2.75°C   **39.** 10 rows   **41.** 12 years, $11 700
43. 490 m   **45.** $S_n = \frac{1}{2}n[2a_1 + (n - 1)d]$ **47.** $a_1 = 1$, $a_n = n$

Exercises 18-2, page 509

1. 45, 15, 5, $\frac{5}{3}$, $\frac{5}{9}$ **3.** 2, 6, 18, 54, 162   **5.** 16   **7.** $\frac{1}{125}$ **9.** $\frac{1}{9}$ **11.** 2×10^6 **13.** $\frac{341}{8}$ **15.** 378
17. $a_6 = 64$, $S_6 = \frac{1365}{16}$ **19.** $a_1 = 16$, $a_5 = 81$ **21.** $a_1 = 1$, $r = 3$ **23.** $n = 5$, $S_5 = \frac{2343}{25}$ **25.** 32

27. 1.4%   **29.** 1.09 mA   **31.** $11 277.89   **33.** 462 cm   **35.** 4   **37.** $671 088.64   **39.** $S_n = \dfrac{a_1 - ra_n}{1 - r}$

Exercises 18-3, page 513

1. 8   **3.** $\frac{25}{4}$ **5.** $\frac{400}{21}$ **7.** 8   **9.** $\dfrac{10\,000}{9999}$ **11.** $\frac{1}{2}(5 + 3\sqrt{3})$ **13.** $\frac{1}{3}$ **15.** $\frac{40}{99}$ **17.** $\frac{2}{11}$ **19.** $\frac{91}{333}$

21. $\frac{11}{30}$ **23.** $\dfrac{100\,741}{999\,000}$ **25.** 350 L   **27.** 346 g

Exercises 18-4, page 517

1. $t^3 + 3t^2 + 3t + 1$ **3.** $16x^4 - 32x^3 + 24x^2 - 8x + 1$ **5.** $32x^5 + 240x^4 + 720x^3 + 1080x^2 + 810x + 243$
7. $64a^6 - 192a^5b^2 + 240a^4b^4 - 160a^3b^6 + 60a^2b^8 - 12ab^{10} + b^{12}$ **9.** $625x^4 - 1500x^3 + 1350x^2 - 540x + 81$
11. $64a^6 + 192a^5 + 240a^4 + 160a^3 + 60a^2 + 12a + 1$ **13.** $x^{10} + 20x^9 + 180x^8 + 960x^7 + \cdots$
15. $128a^7 - 448a^6 + 672a^5 - 560a^4 + \cdots$ **17.** $x^{24} - 6x^{22}y + \frac{33}{2}x^{20}y^2 - \frac{55}{2}x^{18}y^3 + \cdots$
19. $b^{40} + 10b^{37} + \frac{95}{2}b^{34} + \frac{285}{2}b^{31} + \cdots$ **21.** $1 + 8x + 28x^2 + 56x^3 + \cdots$ **23.** $1 + 2x + 3x^2 + 4x^3 + \cdots$
25. $1 + \frac{1}{2}x - \frac{1}{8}x^2 + \frac{1}{16}x^3 - \cdots$ **27.** $\frac{1}{3}[1 + \frac{1}{2}x + \frac{3}{8}x^2 + \frac{5}{16}x^3 + \cdots]$
29. (a) 3.557×10^{14}, (b) 5.109×10^{19}, (c) 8.536×10^{15}, (d) 2.480×10^{96}
31. $n! = n(n - 1)(n - 2)(\cdots)(2)(1) = n \times (n - 1)!$; for $n = 1$, $1! = 1 \times 0! = 1 \times 1 = 1$ **33.** $56a^3b^5$
35. $10{,}264{,}320x^8b^4$ **37.** $V = A(1 - 5r + 10r^2 - 10r^3 + 5r^4 - r^5)$ **39.** $1 - \dfrac{x}{a} + \dfrac{x^3}{2a^3} - \cdots$

Review Exercises for Chapter 18, page 519

1. 81   **3.** 1.28×10^{-6} **5.** $-\frac{119}{2}$ **7.** $\frac{16}{243}$ **9.** $\frac{195}{2}$ **11.** $\frac{1023}{96}$ **13.** 81   **15.** $\frac{9}{16}$ **17.** -1.5
19. -0.25 **21.** 186   **23.** $\frac{455}{2}$ (as), 127 (gs), or 43 (gs)   **25.** 27   **27.** 51   **29.** $\frac{1}{33}$ **31.** $\frac{8}{110}$
33. $x^4 - 8x^3 + 24x^2 - 32x + 16$ **35.** $x^{10} + 5x^8 + 10x^6 + 10x^4 + 5x^2 + 1$
37. $a^{10} + 20a^9b^2 + 180a^8b^4 + 960a^7b^6 + \cdots$ **39.** $p^{18} - \frac{3}{2}p^{16}q + p^{14}q^2 - \frac{7}{18}p^{12}q^3 + \cdots$
41. $1 + 12x + 66x^2 + 220x^3 + \cdots$ **43.** $1 + \frac{1}{2}x^2 - \frac{1}{8}x^4 + \frac{1}{16}x^6 - \cdots$ **45.** $1 - \frac{1}{2}a^2 - \frac{1}{8}a^4 - \frac{1}{16}a^6 - \cdots$
47. $1 + 6x + 24x^2 + 80x^3 + \cdots$ **49.** 1 001 000   **51.** 11th   **53.** 12.6 mm   **55.** 769.4 cm   **57.** $2391.24
59. 1.65×10^{10} cm = 165 000 km   **61.** $47 340.80   **63.** $6.93   **65.** $1 + \frac{1}{2}am^2 + \frac{1}{8}am^4$ **67.** (a) No, (b) Yes

Exercises 19-1, page 528

(*Note:* "Answers" to trigonometric identities are intermediate steps of suggested reductions of the left member.)

1. $1.483 = \dfrac{1}{0.6745}$ **3.** $\left(-\frac{1}{2}\sqrt{3}\right)^2 + \left(-\frac{1}{2}\right)^2 = \frac{3}{4} + \frac{1}{4} = 1$ **5.** $\dfrac{\cos\theta}{\sin\theta}\left(\dfrac{1}{\cos\theta}\right) = \dfrac{1}{\sin\theta}$ **7.** $\dfrac{\sin x}{\dfrac{\sin x}{\cos x}} = \dfrac{\sin x}{1}\left(\dfrac{\cos x}{\sin x}\right)$

9. $\sin y\left(\dfrac{\cos y}{\sin y}\right)$ **11.** $\sin x\left(\dfrac{1}{\cos x}\right)$ **13.** $\csc^2 x(\sin^2 x)$

15. $\sin x(\csc^2 x) = (\sin x)(\csc x)(\csc x) = \sin x\left(\dfrac{1}{\sin x}\right)\csc x$ **17.** $\sin x \csc x - \sin^2 x = 1 - \sin^2 x$

19. $\tan y \cot y + \tan^2 y = 1 + \tan^2 y$ **21.** $\sin x\left(\dfrac{\sin x}{\cos x}\right) + \cos x = \dfrac{\sin^2 x + \cos^2 x}{\cos x} = \dfrac{1}{\cos x}$

23. $\cos\theta\left(\dfrac{\cos\theta}{\sin\theta}\right) + \sin\theta = \dfrac{\cos^2\theta + \sin^2\theta}{\sin\theta} = \dfrac{1}{\sin\theta}$

25. $\sec\theta\left(\dfrac{\sin\theta}{\cos\theta}\right)\csc\theta = \sec\theta\left(\dfrac{1}{\cos\theta}\right)(\sin\theta\,\csc\theta) = \sec\theta(\sec\theta)(1)$

27. $\cot\theta(\sec^2\theta - 1) = \cot\theta\,\tan^2\theta = (\cot\theta\,\tan\theta)\tan\theta$ **29.** $\dfrac{\sin x}{\cos x} + \dfrac{\cos x}{\sin x} = \dfrac{\sin^2 x + \cos^2 x}{\cos x \sin x} = \dfrac{1}{\cos x \sin x}$

31. $(1 - \sin^2 x) - \sin^2 x$ **33.** $\dfrac{\sin x(1 + \cos x)}{1 - \cos^2 x} = \dfrac{1 + \cos x}{\sin x}$

35. $\dfrac{(1/\cos x) + (1/\sin x)}{1 + (\sin x/\cos x)} = \dfrac{(\sin x + \cos x)/\cos x \sin x}{(\cos x + \sin x)/\cos x} = \dfrac{\cos x}{\cos x \sin x}$

37. $\dfrac{\sin^2 x}{\cos^2 x}\cos^2 x + \dfrac{\cos^2 x}{\sin^2 x}\sin^2 x = \sin^2 x + \cos^2 x$ **39.** $\dfrac{\sec\theta}{\dfrac{1}{\sec\theta}} - \dfrac{\tan\theta}{\dfrac{1}{\tan\theta}} = \sec^2\theta - \tan^2\theta$

41. $\dfrac{\sin^2 x + \cos^2 x - 2\cos^2 x}{\sin x \cos x} = \dfrac{\sin^2 x - \cos^2 x}{\sin x \cos x} = \dfrac{\sin x}{\cos x} - \dfrac{\cos x}{\sin x}$ **43.** $\cos^3 x\left(\dfrac{1}{\sin^3 x}\right)\left(\dfrac{\sin^3 x}{\cos^3 x}\right) = 1$

45. $\dfrac{1}{\cos x} + \dfrac{\sin x}{\cos x} + \dfrac{\cos x}{\sin x} = \dfrac{\sin x + \cos^2 x + \sin^2 x}{\sin x \cos x}$ **47.** $\dfrac{\cos\theta + \sin\theta}{1 + \dfrac{\sin\theta}{\cos\theta}} = \dfrac{\cos\theta + \sin\theta}{\dfrac{\cos\theta + \sin\theta}{\cos\theta}}$

49. $\left(\dfrac{\sin x}{\cos x} + \dfrac{\cos x}{\sin x}\right)\sin x \cos x = \dfrac{\sin^2 x \cos x}{\cos x} + \dfrac{\sin x \cos^2 x}{\sin x}$

51. $\dfrac{(\sin^2 x - \cos^2 x)(\sin^2 x + \cos^2 x)}{(1 - \cot^2 x)(1 + \cot^2 x)} = \dfrac{\sin^2 x - \cos^2 x}{[1 - (\cos^2 x/\sin^2 x)]\csc^2 x}$ **53.** $\sec^2 x - 1 + 1 - \tan x + \tan x$

55. Infinite series: $\dfrac{1}{1 - \sin^2 x} = \dfrac{1}{\cos^2 x}$ **57.** $0 = \cos A \cos B \cos C + \sin A \sin B, \cos C = -\dfrac{\sin A \sin B}{\cos A \cos B}$

59. $\ell = a \csc\theta + a \sec\theta = a\left(\dfrac{1}{\sin\theta} + \dfrac{\tan\theta}{\sin\theta}\right)$ **61.** $\sqrt{1 - \cos^2\theta} = \sqrt{\sin^2\theta}$ **63.** $\sqrt{4 + 4\tan^2\theta} = 2\sqrt{1 + \tan^2\theta}$

Exercises 19-2, page 533

1. $\sin 105° = \sin 60° \cos 45° + \cos 60° \sin 45° = \dfrac{\sqrt{3}}{2}\dfrac{\sqrt{2}}{2} + \dfrac{1}{2}\dfrac{\sqrt{2}}{2} = 0.9659$

3. $\cos 15° = \cos(60° - 45°) = \cos 60° \cos 45° + \sin 60° \sin 45°$
$= (\tfrac{1}{2})(\tfrac{1}{2}\sqrt{2}) + (\tfrac{1}{2}\sqrt{3})(\tfrac{1}{2}\sqrt{2}) = \tfrac{1}{4}\sqrt{2} + \tfrac{1}{4}\sqrt{6} = \tfrac{1}{4}(\sqrt{2} + \sqrt{6}) = 0.9659$

5. $-\tfrac{33}{65}$ **7.** $-\tfrac{56}{65}$ **9.** $\sin 3x$ **11.** $\cos x$ **13.** $\cos(2 - x)$ **15.** 0 **17.** 1 **19.** 1

21. $\sin(180° - x) = \sin 180° \cos x - \cos 180° \sin x = (0)\cos x - (-1)\sin x$

23. $\cos(0 - x) = \cos 0 \cos x + \sin 0 \sin x = (1)\cos x + (0)\sin x$

25. $\sin(270° - x) = \sin 270° \cos x - \cos 270° \sin x = (-1)\cos x - 0(\sin x)$

27. $\cos(\tfrac{1}{2}\pi - x) = \cos\tfrac{1}{2}\pi \cos x + \sin\tfrac{1}{2}\pi \sin x = 0(\cos x) + 1(\sin x)$

29. $\cos(30° + x) = \cos 30° \cos x - \sin 30° \sin x = \tfrac{1}{2}\sqrt{3}\cos x - \tfrac{1}{2}\sin x = \tfrac{1}{2}(\sqrt{3}\cos x - \sin x)$

31. $\sin\left(\dfrac{\pi}{4} + x\right) = \sin\dfrac{\pi}{4}\cos x + \cos\dfrac{\pi}{4}\sin x = \tfrac{1}{2}\sqrt{2}\cos x + \tfrac{1}{2}\sqrt{2}\sin x$

33. $(\sin x \cos y + \cos x \sin y)(\sin x \cos y - \cos x \sin y) = \sin^2 x \cos^2 y - \cos^2 x \sin^2 y$
$= \sin^2 x (1 - \sin^2 y) - (1 - \sin^2 x)\sin^2 y$

35. $(\cos\alpha \cos\beta - \sin\alpha \sin\beta) + (\cos\alpha \cos\beta + \sin\alpha \sin\beta)$ **37, 39, 41, 43.** Use the indicated method.

45. $i_0 \sin(\omega t + \alpha) = i_0 (\sin\omega t \cos\alpha + \cos\omega t \sin\alpha)$

47. $\tan\alpha (R + \cos\beta) = \sin\beta, R = \dfrac{\sin\beta - \tan\alpha \cos\beta}{\tan\alpha} = \dfrac{\sin\beta \cos\alpha - \cos\beta \sin\alpha}{\cos\alpha \tan\alpha}$

Exercises 19-3, page 538

1. $\sin 60° = \sin 2(30°) = 2 \sin 30° \cos 30° = 2(\tfrac{1}{2})(\tfrac{1}{2}\sqrt{3}) = \tfrac{1}{2}\sqrt{3}$

3. $\cos 120° = \cos 2(60°) = \cos^2 60° - \sin^2 60° = (\tfrac{1}{2})^2 - (\tfrac{1}{2}\sqrt{3})^2 = -\tfrac{1}{2}$

5. $\sin 258° = 2 \sin 129° \cos 129° = -0.978\ 147\ 6$ **7.** $\cos 96° = \cos^2 48° - \sin^2 48° = -0.104\ 528\ 5$ **9.** $\tfrac{24}{25}$

11. $\tfrac{3}{5}$ **13.** $2 \sin 8x$ **15.** $\cos 8x$ **17.** $\cos x$ **19.** $-2 \cos 4x$ **21.** $\cos^2 \alpha - (1 - \cos^2 \alpha)$

23. $\dfrac{\cos x - (\sin x/\cos x)\sin x}{1/\cos x} = \cos^2 x - \sin^2 x$ **25.** $(\cos^2 x - \sin^2 x)(\cos^2 x + \sin^2 x) = (\cos^2 x - \sin^2 x)(1)$

27. $\dfrac{2 \sin 2\theta \cos 2\theta}{\sin 2\theta}$ **29.** $\dfrac{2 \sin \theta \cos \theta}{1 + 2 \cos^2 \theta - 1} = \dfrac{\sin \theta}{\cos \theta}$ **31.** $\dfrac{1}{\sec^2 x} - \dfrac{\tan^2 x}{\sec^2 x} = \cos^2 x - \dfrac{\sin^2 x}{\cos^2 x \sec^2 x}$

33. $\dfrac{2 \tan x}{\sin 2x} = \dfrac{2(\sin x/\cos x)}{2 \sin x \cos x} = \dfrac{1}{\cos^2 x}$ **35.** $\dfrac{\sin 3x \cos x - \cos 3x \sin x}{\sin x \cos x} = \dfrac{\sin 2x}{\frac{1}{2} \sin 2x}$

37. $\sin(2x + x) = \sin 2x \cos x + \cos 2x \sin x = (2 \sin x \cos x)(\cos x) + (\cos^2 x - \sin^2 x)\sin x$

39. Use the indicated method.   **41.** $R = v\left(\dfrac{2v \sin \alpha}{g}\right)\cos \alpha = \dfrac{v^2(2 \sin \alpha \cos \alpha)}{g}$

43. $vi \sin \omega t \sin\left(\omega t - \dfrac{\pi}{2}\right) = vi \sin \omega t\left(\sin \omega t \cos \dfrac{\pi}{2} - \cos \omega t \sin \dfrac{\pi}{2}\right) = vi \sin \omega t[-(\cos \omega t)(1)] = -\dfrac{1}{2} vi(2 \sin \omega t \cos \omega t)$

Exercises 19-4, page 542

1. $\cos 15° = \cos \dfrac{1}{2}(30°) = \sqrt{\dfrac{1 + \cos 30°}{2}} = \sqrt{\dfrac{1.8660}{2}} = 0.9659$

3. $\sin 75° = \sin \dfrac{1}{2}(150°) = \sqrt{\dfrac{1 - \cos 150°}{2}} = \sqrt{\dfrac{1.8660}{2}} = 0.9659$ **5.** $\sin 118° = 0.882\ 947\ 6$

7. $\sqrt{2\left(\dfrac{1 + \cos 164°}{2}\right)} = \sqrt{2} \cos 82° = 0.196\ 820\ 5$ **9.** $\sin 3x$ **11.** $4 \cos 2x$ **13.** $\tfrac{1}{26}\sqrt{26}$ **15.** $\tfrac{1}{10}\sqrt{2}$

17. $\pm\sqrt{\dfrac{2}{1 - \cos \alpha}}$ **19.** $\tan \dfrac{1}{2}\alpha = \dfrac{1 - \cos \alpha}{\sin \alpha} = \dfrac{\sin \alpha}{1 + \cos \alpha}$ **21.** $\dfrac{1 - \cos \alpha}{2 \sin \frac{1}{2}\alpha} = \dfrac{1 - \cos \alpha}{2\sqrt{\frac{1}{2}(1 - \cos \alpha)}} = \sqrt{\dfrac{1 - \cos \alpha}{2}}$

23. $2\left(\dfrac{1 - \cos x}{2}\right) + \cos x$ **25.** $\sqrt{\dfrac{(1 + \cos \theta)(1 - \cos \theta)}{2(1 - \cos \theta)}} = \dfrac{\sin \theta}{\sqrt{4\left(\dfrac{1 - \cos \theta}{2}\right)}}$ **27.** $2\left(\dfrac{1 - \cos \alpha}{2}\right) - \left(\dfrac{1 + \cos \alpha}{2}\right)$

29. $\sin \omega t = \pm\sqrt{\dfrac{1 - \cos 2\omega t}{2}}$, $\sin^2 \omega t = \tfrac{1}{2}(1 - \cos 2\omega t)$

31. $n = \left(\sqrt{\dfrac{1 - \cos (A + \phi)}{2}}\right)\bigg/\left(\sqrt{\dfrac{1 - \cos A}{2}}\right) = \sqrt{\dfrac{1 - \cos (A + \phi)}{1 - \cos A}}$

Exercises 19-5, page 546

1. $\dfrac{\pi}{2}$ **3.** $\dfrac{3\pi}{4}, \dfrac{7\pi}{4}$ **5.** π **7.** $0.9273, 2.214$ **9.** $\dfrac{\pi}{3}, \dfrac{2\pi}{3}, \dfrac{4\pi}{3}, \dfrac{5\pi}{3}$ **11.** $\dfrac{\pi}{3}, \dfrac{2\pi}{3}, \dfrac{4\pi}{3}, \dfrac{5\pi}{3}$ **13.** $0, \dfrac{\pi}{6}, \dfrac{5\pi}{6}, \pi$

15. $\dfrac{\pi}{12}, \dfrac{\pi}{4}, \dfrac{5\pi}{12}, \dfrac{3\pi}{4}, \dfrac{13\pi}{12}, \dfrac{5\pi}{4}, \dfrac{17\pi}{12}, \dfrac{7\pi}{4}$ **17.** $\dfrac{\pi}{2}, \dfrac{3\pi}{2}$ **19.** $0, \dfrac{\pi}{3}, \pi, \dfrac{5\pi}{3}$ **21.** $\dfrac{\pi}{4}, \dfrac{3\pi}{4}, \dfrac{5\pi}{4}, \dfrac{7\pi}{4}$ **23.** $3.569, 5.856$

25. $0.2618, 1.309, 3.403, 4.451$ **27.** $0.7854, 1.249, 3.927, 4.391$ **29.** $\dfrac{3\pi}{8}, \dfrac{7\pi}{8}, \dfrac{11\pi}{8}, \dfrac{15\pi}{8}$ **31.** $0, \pi$

33. 6.56×10^{-4} **35.** $10.2\ \text{s}, 15.7\ \text{s}, 21.2\ \text{s}, 47.1\ \text{s}$ **37.** $0.3398, 2.802$ **39.** $0.8861, 2.256, 4.028, 5.397$

41. $-0.95, 0.00, 0.95$ **43.** 1.08

Exercises 19-6 page 552

1. y is the angle whose tangent is x.   **3.** y is the angle whose cotangent is $3x$.

5. y is twice the angle whose sine is x.   **7.** y is 5 times the angle whose cosine is $2x$.   **9.** $\dfrac{\pi}{3}$ **11.** 0 **13.** $-\dfrac{\pi}{3}$

15. $\dfrac{\pi}{3}$ **17.** $\dfrac{\pi}{6}$ **19.** $-\dfrac{\pi}{4}$ **21.** $\dfrac{\pi}{4}$ **23.** $-\dfrac{\pi}{3}$ **25.** $\dfrac{3\pi}{4}$ **27.** $\tfrac{1}{2}\sqrt{3}$ **29.** $\tfrac{1}{2}\sqrt{2}$ **31.** -1 **33.** -1.3090

35. -0.9838 **37.** 1.4413 **39.** 1.4503 **41.** -1.2389 **43.** -0.2239 **45.** $x = \tfrac{1}{3} \text{Arcsin } y$

47. $x = 4 \tan y$ **49.** $x = \tfrac{1}{3} \text{Arcsec } (y - 1)$ **51.** $x = 1 - \cos (1 - y)$ **53.** $\dfrac{x}{\sqrt{1 - x^2}}$ **55.** $\dfrac{1}{x}$ **57.** $\dfrac{3x}{\sqrt{9x^2 - 1}}$

59. $2x\sqrt{1-x^2}$ **61.** $t = \dfrac{1}{2\omega}\,\text{Arccos}\,\dfrac{y}{A} - \dfrac{\phi}{\omega}$ **63.** $t = \dfrac{1}{\omega}\left(\text{Arcsin}\,\dfrac{i}{I_m} - \alpha - \phi\right)$

65. $\sin(\text{Arcsin}\,\tfrac{3}{5} + \text{Arcsin}\,\tfrac{5}{13}) = \tfrac{3}{5}\tfrac{12}{13} + \tfrac{4}{5}\tfrac{5}{13} = \tfrac{56}{65}$ **67.** $\dfrac{\pi}{6} + \dfrac{\pi}{3} = \dfrac{\pi}{2}$ **69.** $\text{Arcsin}\left(\dfrac{a}{c}\right)$

71. Let y = height of cliff; $\tan\alpha = \dfrac{h+y}{d}$, $\tan\beta = \dfrac{y}{d}$; $\tan\alpha = \dfrac{h + d\tan\beta}{d}$

Review Exercises for Chapter 19, page 555

1. $\sin(90° + 30°) = \sin 90°\cos 30° + \cos 90°\sin 30° = (1)(\tfrac{1}{2}\sqrt{3}) + (0)(\tfrac{1}{2}) = \tfrac{1}{2}\sqrt{3}$
3. $\sin(180° - 45°) = \sin 180°\cos 45° - \cos 180°\sin 45° = 0(\tfrac{1}{2}\sqrt{2}) - (-1)(\tfrac{1}{2}\sqrt{2}) = \tfrac{1}{2}\sqrt{2}$
5. $\cos 2(90°) = \cos^2 90° - \sin^2 90° = 0 - 1 = -1$ **7.** $\sin\tfrac{1}{2}(90°) = \sqrt{\tfrac{1}{2}(1 - \cos 90°)} = \sqrt{\tfrac{1}{2}(1-0)} = \tfrac{1}{2}\sqrt{2}$
9. $\sin 52° = 0.788\,010\,8$ **11.** $\sin 92° = 0.999\,390\,8$ **13.** $\cos 6° = 0.994\,521\,9$ **15.** $\cos 164° = -0.961\,261\,7$

17. $\sin 5x$ **19.** $4\sin 12x$ **21.** $2\cos 12x$ **23.** $2\cos x$ **25.** $-\dfrac{\pi}{2}$ **27.** 0.2619 **29.** $-\tfrac{1}{3}\sqrt{3}$ **31.** 0

33. $\dfrac{\dfrac{1}{\cos y}}{\dfrac{1}{\sin y}} = \dfrac{1}{\cos y}\dfrac{\sin y}{1}$ **35.** $\sin x\csc x - \sin^2 x = 1 - \sin^2 x$ **37.** $\dfrac{1 - \sin^2\theta}{\sin\theta} = \dfrac{\cos^2\theta}{\sin\theta}$

39. $\cos\theta\left(\dfrac{\cos\theta}{\sin\theta}\right) + \sin\theta = \dfrac{\cos^2\theta + \sin^2\theta}{\sin\theta}$ **41.** $\dfrac{(\sec^2 x - 1)(\sec^2 x + 1)}{\tan^2 x} = \sec^2 x + 1$

43. $2\left(\dfrac{1}{\sin 2x}\right)\left(\dfrac{\cos x}{\sin x}\right) = 2\left(\dfrac{1}{2\sin x\cos x}\right)\left(\dfrac{\cos x}{\sin x}\right) = \dfrac{1}{\sin^2 x}$ **45.** $\dfrac{\cos^2\theta}{\sin^2\theta}$ **47.** $\dfrac{(\cos^2\theta - \sin^2\theta)}{\cos^2\theta} = 1 - \dfrac{\sin^2\theta}{\cos^2\theta}$

49. $\dfrac{1}{2}\left(2\sin\dfrac{\theta}{2}\cos\dfrac{\theta}{2}\right)$ **51.** $\dfrac{1}{\cos x} + \dfrac{\sin x}{\cos x} = \dfrac{(1 + \sin x)(1 - \sin x)}{\cos x(1 - \sin x)} = \dfrac{1 - \sin^2 x}{\cos x(1 - \sin x)}$ **53.** $\cos[(x - y) + y]$

55. $\sin 4x(\cos 4x)$ **57.** $\dfrac{\sin x}{\dfrac{1}{\sin x} - \dfrac{\cos x}{\sin x}} = \dfrac{\sin^2 x}{1 - \cos x} = \dfrac{1 - \cos^2 x}{1 - \cos x}$

59. $\dfrac{\sin x\cos y + \cos x\sin y + \sin x\cos y - \cos x\sin y}{\cos x\cos y - \sin x\sin y + \cos x\cos y + \sin x\sin y} = \dfrac{2\sin x\cos y}{2\cos x\cos y}$ **61.** $x = \dfrac{1}{2}\text{Arccos}\,\dfrac{1}{2}y$

63. $x = \dfrac{1}{5}\sin\dfrac{1}{3}\left(\dfrac{1}{4}\pi - y\right)$ **65.** $1.2925,\ 4.4341$ **67.** $\dfrac{\pi}{6}, \dfrac{5\pi}{6}, \dfrac{7\pi}{6}, \dfrac{11\pi}{6}$ **69.** $0, \dfrac{\pi}{2}, \pi, \dfrac{3\pi}{2}$ **71.** $\dfrac{\pi}{6}, \dfrac{5\pi}{6}, \dfrac{7\pi}{6}, \dfrac{11\pi}{6}$

73. $0, \pi$ **75.** 0 **77.** $\dfrac{1}{x}$ **79.** $2x\sqrt{1-x^2}$ **81.** $2\sqrt{1 - \cos^2\theta}$ **83.** $\dfrac{\tan\theta}{\sqrt{1 + \tan^2\theta}} = \dfrac{\tan\theta}{\sec\theta}$

85. $R = \sqrt{(A\cos\theta - B\sin\theta)^2 + (A\sin\theta + B\cos\theta)^2} = \sqrt{A^2(\cos^2\theta + \sin^2\theta) + B^2(\sin^2\theta + \cos^2\theta)}$

87. $\cos\alpha = \dfrac{A}{C}$, $\sin\alpha = \dfrac{B}{C}$:   **89.** $\theta = \alpha + R\sin\omega t$

$A\sin 2t + B\cos 2t = C\left(\dfrac{A}{C}\sin 2t + \dfrac{B}{C}\cos 2t\right)$
$= C(\cos\alpha\sin 2t + \sin\alpha\cos 2t)$

91. $1 - (1 - 2\sin^2\phi) - (2\sin\phi\cos\phi)\tan\alpha = 2\sin^2\phi - 2\sin\phi\cos\phi\tan\alpha$ **93.** $83.7°$ **95.** $54.7°$

Exercises 20-1, page 563

1. $2\sqrt{29}$ **3.** 3 **5.** 55 **7.** 7 **9.** 2.86 **11.** $\tfrac{5}{2}$ **13.** Undefined   **15.** $-\tfrac{3}{4}$ **17.** 0 **19.** 0.747
21. $\tfrac{1}{3}\sqrt{3}$ **23.** $-\tfrac{1}{3}\sqrt{3}$ **25.** $20.0°$ **27.** $98.50°$ **29.** Parallel   **31.** Perpendicular   **33.** $8, -2$
35. -3 **37.** Two sides equal $2\sqrt{10}$ **39.** $m_1 = \tfrac{5}{12}, m_2 = \tfrac{4}{3}$ **41.** 10 **43.** $4\sqrt{10} + 4\sqrt{2} = 18.3$ **45.** $(1, 5)$
47. $(-3, 4)$

Exercises 20-2, page 569

1. $4x - y + 20 = 0$ **3.** $7x - 2y - 24 = 0$ **5.** $x - y + 2 = 0$ **7.** $y = -3$ **9.** $x = -3$

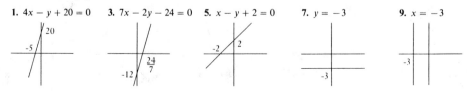

11. $3x - 2y - 12 = 0$ **13.** $x + 3y + 5 = 0$ **15.** $x + 2y - 4 = 0$ **17.** $4x + 3y + 6 = 0$

19. $2x + 5y + 14 = 0$ **21.**   **23.**

25. $y = \frac{3}{2}x - \frac{1}{2}; m = \frac{3}{2}, (0, -\frac{1}{2})$
27. $y = \frac{5}{2}x + \frac{5}{2}; m = \frac{5}{2}, (0, \frac{5}{2})$
29. -2 **31.** 1   **33.** $m_1 = m_2 = \frac{3}{2}$
35. $m_1 = 2, m_2 = -\frac{1}{2}$
37. $x + 2y - 4 = 0$

39. $3x + y - 18 = 0$ **41.** $v = 3.35 + 1.45t$ **43.** $\ell = 2c + 8$ **45.** $5x + 6y = 1220$ **47.** $y = 150 - 0.80x$

49. $y = 10^{-5}(2.4 - 5.6x)$ **51.** $n = \frac{7}{6}t + 10$; at 6:30, $n = 10$; at 8:30, $n = 150$
53.   **55.**   **57.**   **59.**

    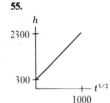   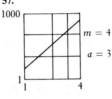   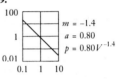

Exercises 20-3, page 574

1. $(2, 1), r = 5$ **3.** $(-1, 0), r = 2$ **5.** $x^2 + y^2 = 9$
7. $(x - 2)^2 + (y - 2)^2 = 16$, or $x^2 + y^2 - 4x - 4y - 8 = 0$
9. $(x + 2)^2 + (y - 5)^2 = 5$, or $x^2 + y^2 + 4x - 10y + 24 = 0$
11. $(x - 12)^2 + (y + 15)^2 = 324$, or $x^2 + y^2 - 24x + 30y + 45 = 0$
13. $(x - 2)^2 + (y - 1)^2 = 8$, or $x^2 + y^2 - 4x - 2y - 3 = 0$
15. $(x + 3)^2 + (y - 5)^2 = 25$, or $x^2 + y^2 + 6x - 10y + 9 = 0$
17. $(x - 2)^2 + (y - 2)^2 = 4$, or $x^2 + y^2 - 4x - 4y + 4 = 0$
19. $(x - 2)^2 + (y - 5)^2 = 25$, or $x^2 + y^2 - 4x - 10y + 4 = 0$; and
$(x + 2)^2 + (y + 5)^2 = 25$, or $x^2 + y^2 + 4x + 10y + 4 = 0$

21. $(0, 3)$,   **23.** $(-1, 5)$,   **25.** $(0, 0)$,   **27.** $(1, 0)$
$r = 2$ $r = \frac{9}{2}$ $r = 5$ $r = 3$

        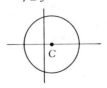

29. $(-4, 5)$,   **31.** $(1, 2)$,   **33.** Symmetrical to both axes and origin
$r = 7$ $r = \frac{1}{2}\sqrt{22}$ **35.** Symmetrical to y-axis   **37.** $(7, 0), (-1, 0)$
                        39. $3x^2 + 3y^2 + 4x + 8y - 20 = 0$, circle

    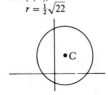

41. $x^2 + y^2 = 0.0100$ **43.** $(x - 500 \times 10^{-6})^2 + y^2 = 0.16 \times 10^{-6}$

Exercises 20-4, page 579

1. $F(1, 0)$, $x = -1$ **3.** $F(-1, 0)$, $x = 1$ **5.** $F(0, 2)$, $y = -2$ **7.** $F(0, -1)$, $y = 1$

9. $F(\frac{1}{2}, 0)$, $x = -\frac{1}{2}$ **11.** $F(0, \frac{1}{4})$, $y = -\frac{1}{4}$ **13.** $y^2 = 12x$ **15.** $x^2 = 16y$ **17.** $x^2 = 4y$
                                                                                                       19. $x^2 = \frac{1}{8}y$

21. $y^2 - 2y - 12x + 37 = 0$ **23.** $x^2 - 2x + 8y - 23 = 0$ **25.** $4p$ **27.** H

29. 0.919 m   **31.** 1.48 m   **33.**

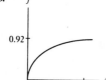

35. $y^2 = 8x$ or $x^2 = 8y$ with vertex midway between island and shore

Exercises 20-5, page 585

1. $V(2, 0)$, $V(-2, 0)$,
     $F(\sqrt{3}, 0)$, $F(-\sqrt{3}, 0)$

3. $V(0, 6)$, $V(0, -6)$,
     $F(0, \sqrt{11})$, $F(0, -\sqrt{11})$

5. $V(3, 0)$, $V(-3, 0)$,
     $F(\sqrt{5}, 0)$, $F(-\sqrt{5}, 0)$

   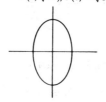   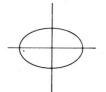

7. $V(0, 7)$, $V(0, -7)$,
$F(0, \sqrt{45})$, $F(0, -\sqrt{45})$,

9. $V(0, 4)$, $V(0, -4)$,
$F(0, \sqrt{14})$, $F(0, -\sqrt{14})$

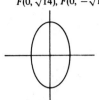

11. $V\left(\frac{5}{2}, 0\right)$, $V\left(-\frac{5}{2}, 0\right)$,
$F\left(\frac{\sqrt{21}}{2}, 0\right)$, $F\left(\frac{-\sqrt{21}}{2}, 0\right)$

13. $\dfrac{x^2}{225} + \dfrac{y^2}{144} = 1$, or $144x^2 + 225y^2 = 32\,400$ **15.** $\dfrac{y^2}{9} + \dfrac{x^2}{5} = 1$, or $9x^2 + 5y^2 = 45$

17. $\dfrac{x^2}{64} + \dfrac{15y^2}{144} = 1$, or $3x^2 + 20y^2 = 192$ **19.** $\dfrac{x^2}{5} + \dfrac{y^2}{20} = 1$, or $4x^2 + y^2 = 20$

21. $16x^2 + 25y^2 - 32x - 50y - 359 = 0$ **23.** $9x^2 + 5y^2 - 18x - 20y - 16 = 0$

25. $2x^2 + 3y^2 - 8x - 4 = 2x^2 + 3(-y)^2 - 8x - 4$ **27.** $\frac{2}{3}\sqrt{2} = 0.943$ **29.** (a) 5.8 m, (b) 5.0 m

31. 27.5 m **33.** $7x^2 + 16y^2 = 112$ **35.** 46.2 m³

Exercises 20-6, page 592

1. $V(5, 0)$, $V(-5, 0)$,
$F(13, 0)$, $F(-13, 0)$

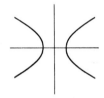

3. $V(0, 3)$, $V(0, -3)$,
$F(0, \sqrt{10})$, $F(0, -\sqrt{10})$

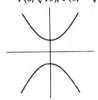

5. $V(1, 0)$, $V(-1, 0)$,
$F(\sqrt{5}, 0)$, $F(-\sqrt{5}, 0)$

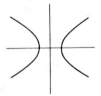

7. $V(0, \sqrt{5})$, $V(0, -\sqrt{5})$,
$F(0, \sqrt{7})$, $F(0, -\sqrt{7})$

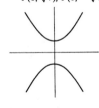

9. $V(0, 2)$, $V(0, -2)$,
$F(0, \sqrt{5})$, $F(0, -\sqrt{5})$

11. $V(2, 0)$, $V(-2, 0)$,
$F(\frac{2}{3}\sqrt{13}, 0)$, $F(-\frac{2}{3}\sqrt{13}, 0)$

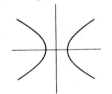

13. $\dfrac{x^2}{9} - \dfrac{y^2}{16} = 1$, or $16x^2 - 9y^2 = 144$ **15.** $\dfrac{y^2}{100} - \dfrac{x^2}{36} = 1$, or $9y^2 - 25x^2 = 900$

17. $\dfrac{x^2}{1} - \dfrac{y^2}{3} = 1$, or $3x^2 - y^2 = 3$ **19.** $\dfrac{x^2}{5} - \dfrac{y^2}{4} = 1$, or $4x^2 - 5y^2 = 20$

21.

23.

25. $9x^2 - 16y^2 - 108x + 64y + 116 = 0$
27. $9x^2 - y^2 - 36x + 27 = 0$ **29.** $x^2 - 2y^2 = 2$

31. $\ell^2 - x^2 = 2000^2$ **33.** $i = 6.00/R$ **35.** Dist. from rifle to P − dist. from target to P = constant (related to dist. from rifle to target).

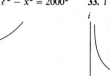

Exercises 20-7, page 596

1. Parabola, $(-1, 2)$ **3.** Hyperbola, $(1, 2)$ **5.** Ellipse, $(-1, 0)$ **7.** Parabola, $(-3, 1)$

9. $(y - 3)^2 = 16(x + 1)$, or $y^2 - 6y - 16x - 7 = 0$ **11.** $(x + 3)^2 = 4(y - 2)$, or $x^2 + 6x - 4y + 17 = 0$

13. $\dfrac{(x + 2)^2}{25} + \dfrac{(y - 2)^2}{16} = 1$, or $16x^2 + 25y^2 + 64x - 100y - 236 = 0$

15. $\dfrac{(y - 1)^2}{16} + \dfrac{(x + 2)^2}{4} = 1$, or $4x^2 + y^2 + 16x - 2y + 1 = 0$

17. $\dfrac{(y - 2)^2}{1} - \dfrac{(x + 1)^2}{3} = 1$, or $x^2 - 3y^2 + 2x + 12y - 8 = 0$

19. $\dfrac{(x + 1)^2}{9} - \dfrac{(y - 1)^2}{16} = 1$, or $16x^2 - 9y^2 + 32x + 18y - 137 = 0$

21. Parabola, $(-1, -1)$ **23.** Ellipse, $(-3, 0)$ **25.** Hyperbola, $(0, 4)$ **27.** Parabola, $(1, 0)$

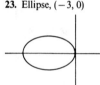   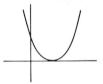

29. $x^2 - y^2 + 4x - 2y - 22 = 0$ **31.** $y^2 + 4x - 4 = 0$ **33.** $(x - 28)^2 = \dfrac{28^2}{18}(y - 18)$

35. $\dfrac{x^2}{9.0} + \dfrac{y^2}{16} = 1$, $\dfrac{(x - 7.0)^2}{16} + \dfrac{y^2}{9.0} = 1$

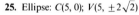

Exercises 20-8, page 600

1. Ellipse   **3.** Hyperbola   **5.** Circle   **7.** Parabola   **9.** Hyperbola   **11.** Circle   **13.** Parabola
15. Hyperbola   **17.** Ellipse   **19.** Ellipse
21. Parabola; $V(-4, 0)$; $F(-4, 2)$ **23.** Hyperbola; $C(1, -2)$;   **25.** Ellipse: $C(5, 0)$; $V(5, \pm2\sqrt{2})$
$V(1, -2 \pm \sqrt{2})$

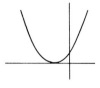   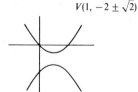

27. Parabola; $V(-\frac{1}{2}, \frac{5}{2})$; $F(\frac{1}{2}, \frac{5}{2})$

29. (a) Circle, (b) hyperbola, (c) ellipse
31. The origin   **33.** Parabola
35. Circle if light beam is perpendicular to floor; otherwise an ellipse

Exercise 20-9, page 604

1.   **3.**   **5.**   **7.**   **9.**   **11.**

   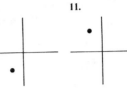

13. $\left(2, \dfrac{\pi}{6}\right)$ **15.** $\left(1, \dfrac{7\pi}{6}\right)$ **17.** $(-4, -4\sqrt{3})$ **19.** $(2.77, -1.15)$ **21.** $r = 3 \sec \theta$ **23.** $r = a$

25. $r = 4 \cot \theta \csc \theta$ **27.** $r^2 = \dfrac{4}{1 + 3\sin^2 \theta}$ **29.** $x^2 + y^2 - y = 0$ **31.** $x = 4$

33. $x^4 + y^4 - 4x^3 + 2x^2y^2 - 4xy^2 - 4y^2 = 0$ **35.** $(x^2 + y^2)^2 = 2xy$ **37.** $B_x = -\dfrac{k \sin \theta}{r}, B_y = \dfrac{k \cos \theta}{r}$

39. $x^4 + y^4 + 2x^2y^2 + 2x^2y + 2y^3 - 9x^2 - 8y^2 = 0$

Exercises 20-10, page 607

1.   **3.**   **5.**   **7.**   **9.**

11.   **13.**   **15.**   **17.**   **19.**

21.   **23.**   **25.**   **27.**   **29.**

31.

Review Exercises for Chapter 20, page 609

1. $4x - y - 11 = 0$

3. $2x + 3y + 3 = 0$

5. $x^2 + y^2 - 2x + 4y - 5 = 0$

7. $y^2 = 12x$

9. $9x^2 + 25y^2 = 900$

11. $144y^2 - 169x^2 = 24{,}336$

13. $(-3, 0), r = 4$

15. $(0, -5), y = 5$

17. $V(0, 4), V(0, -4)$
$F(0, \sqrt{15}), F(0, -\sqrt{15})$

19. $V(2, 0), V(-2, 0)$
$F(\frac{2}{5}\sqrt{35}, 0), F(-\frac{2}{5}\sqrt{35}, 0)$

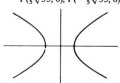

21. $V(4, -8), \ F(4, -7)$

23. $(2, -1)$

25.

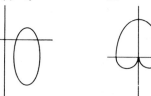

27.

29.

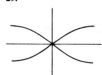

31.

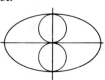

33. $\theta = \text{Arctan } 2 = 1.11$ **35.** $r^2 \cos 2\theta = 16$ **37.** $(x^2 + y^2)^3 = 16x^2y^2$ **39.** $3x^2 + 4y^2 - 8x - 16 = 0$

41. 4   **43.** 2   **45.** $(1.90, 1.55), (1.90, -1.55), (-1.90, 1.55), (-1.90, -1.55)$

47. $m_1 = -\frac{12}{5}, m_2 = \frac{5}{12}; d_1^2 = 169, d_2^2 = 169, d_3^2 = 338$ **49.** $x^2 - 6x - 8y + 1 = 0$

51. $R_T = R + 2.5$ **53.** $y = 25 - \frac{5}{3}x$ **55.** $y = 100.5T - 10\,050$ **57.** $x^2 + y^2 = 0.513$

59. $y^2 = 32x$

61. $A = 300w - w^2$

63.

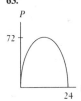

65. $\dfrac{(d-10)^2}{100} + \dfrac{f^2}{1} = 1$

67. $\dfrac{x^2}{3.80 \times 10^6} + \dfrac{y^2}{3.78 \times 10^6} = 1$

69. 11.3 m   **71.** 88.7 cm

73.

75. $(a^2 - b^2)x^2 + a^2 y^2 + 2bx - 1 = 0$;
$a = b$, parabola; $a^2 > b^2$, ellipse; $a^2 < b^2$, hyperbola

Exercises 21-1, page 617

1.

No.	2	3	4	5	6	7
Freq.	1	3	4	2	3	2

3.

No.	0.45	0.46	0.47	0.48	0.49	0.50	0.51	0.52	0.53	0.54	0.55	0.56	0.57
Freq.	1	1	1	2	2	0	1	0	1	0	2	0	1

5.

Int.	2–3	4–5	6–7
Freq.	4	6	5

7.

Int.	0.43–0.45	0.46–0.48	0.49–0.51	0.52–0.54	0.55–0.57
Freq.	1	4	3	1	3

9.

11.

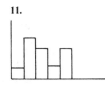

13.

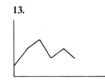

15.

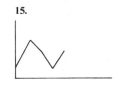

17.

No. inst.	18	19	20	21	22	23	24	25
No.	1	3	2	4	3	1	0	1

19.

21.

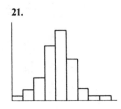

23.

25.

27.

29.

31.

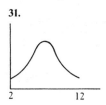

Exercises 21-2, page 622

1. 4   **3.** 0.49   **5.** 4.6   **7.** 0.503   **9.** 4   **11.** 0.48, 0.49, 0.55   **13.** 21   **15.** 21   **17.** 2.248 s
19. 57 m   **21.** 4.237 mR   **23.** 4.36 mR   **25.** 31 h   **27.** 0.005 95 mm   **29.** \$275, \$300
31. 862 kW·h   **33.** 4.5   **35.** 0.51

Exercises 21-3, page 628

1. 1.50   **3.** 0.037   **5.** 1.50   **7.** 0.037   **9.** 1.7   **11.** 2.5 h   **13.** 0.014 s   **15.** 0.000 22 mm   **17.** 60%
19. 58%   **21.** 60%   **23.** 76%

Exercises 21-4, page 632

1. $y = 1.0x - 2.6$

3. $y = -1.77x + 191$

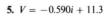
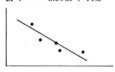

5. $V = -0.590i + 11.3$

7. $h = 2.24x + 5.2$

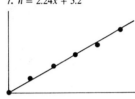

9. $p = -2.66x + 4364$

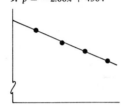

11. $V = 4.32 \times 10^{-15} f - 2.03$
$f_0 = 0.470$ PHz

13. 0.985   **15.** -0.901

Exercises 21-5, page 637

1. $y = 1.97x^2 + 4.8$

3. $y = 10.9/x$

5. $y = 5.97t^2 + 0.38$

7. $p = 2030e^{0.01T} - 1640$

9. $P = \dfrac{1343}{S}$

11. $y = 6.20e^{-t} - 0.05$

Review Exercises for Chapter 21, page 639

1. 3.6   **3.** 0.77   **5.**

Int.	101–103	104–106	107–109	110–112	113–115
Freq.	5	4	3	3	5

7. 106   **9.** 4.6   **11.**

13. 0.264 Pa·s   **15.** 0.014 Pa·s   **17.**

19. 697 W   **21.** 700 W   **23.** 17 W   **25.** 4

27.

29. $R = 0.0983T + 25.0$

31. $s = 0.123t + 0.887$
33. $s = -4.90t^2 + 3000$
35. Divide numerator and denominator by n^2.

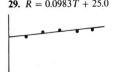

Exercises 22-1, page 649

1. Cont. all x **3.** Not cont. $x = -3$, div. by zero   **5.** Cont. $x > 0$ **7.** Cont. all x
9. Not cont. $x = 1$, small change   **11.** Cont. $x < 2$ **13.** Not cont. $x = 2$, small change
15. Cont. all x

17.

x	2.500	2.900	2.990	2.999	3.001	3.010	3.100	3.500
$f(x)$	5.500	6.700	6.970	6.997	7.003	7.030	7.300	8.500

$\lim\limits_{x \to 3} f(x) = 7$

19.

x	0.900	0.990	0.999	1.001	1.010	1.100
$f(x)$	1.7100	1.9701	1.9970	2.0030	2.0301	2.3100

$\lim\limits_{x \to 1} f(x) = 2$

21.

x	1.900	1.990	1.999	2.001	2.010	2.100
$f(x)$	-0.2516	-0.2502	$-0.250\,02$	$-0.249\,98$	-0.2498	-0.2485

$\lim\limits_{x \to 2} f(x) = -0.25$

23.

x	10	100	1000
$f(x)$	0.4468	0.4044	0.4004

$\lim\limits_{x \to \infty} f(x) = 0.4$ **25.** 7   **27.** 1   **29.** 1

31. -2 **33.** 2   **35.** 2   **37.** Does not exist   **39.** 0   **41.** 3   **43.** 0

45.

x	-0.1	-0.01	-0.001	0.001	0.01	0.1
$f(x)$	-3.1	-3.01	-3.001	-2.999	-2.99	-2.9

$\lim\limits_{x \to 0} f(x) = -3$

47.

x	10	100	1000
$f(x)$	2.1649	2.0106	2.0010

$\lim\limits_{x \to \infty} f(x) = 2$

49. 3 cm/s   **51.** 34.9°C, 0°C   **53.** e **55.** 2

Exercises 22-2, page 655

1. (Slopes) 3.5, 3.9, 3.99, 3.999; $m = 4$

3. (Slopes) $-3.5, -3.9, -3.99, -3.999$; $m = -4$

5. 4   **7.** -4

9. $m_{\text{tan}} = 2x_1$; 4, -2

11. $m_{\text{tan}} = 2x_1 + 2$; $-4, 4$

13. $m_{\text{tan}} = 2x_1 + 4$; $-2, 8$

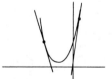

15. $m_{\text{tan}} = 6 - 2x_1$; 10, 0

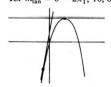

17. $m_{\text{tan}} = 3x_1^2 - 2$; 1, -2, 1

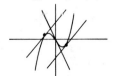

19. $m_{\text{tan}} = 4x_1^3$; 0, 0.5, 4

21. $\dfrac{\Delta y}{\Delta x} = 4.1,\ m_{tan} = 4$ **23.** $\dfrac{\Delta y}{\Delta x} = -12.61,\ m_{tan} = -12$

Exercises 22-3, page 660

1. 3   **3.** -2 **5.** $2x$ **7.** $10x$ **9.** $2x - 7$ **11.** $8 - 4x$ **13.** $3x^2 + 4$ **15.** $-\dfrac{1}{(x+2)^2}$

17. $1 - \dfrac{1}{x^2}$ **19.** $-\dfrac{4}{x^3}$ **21.** $4x^3 + 3x^2 + 2x + 1$ **23.** $4x^3 + \dfrac{2}{x^2}$ **25.** $6x - 2;\ -8$ **27.** $\dfrac{-6}{(x+3)^2};\ -\dfrac{1}{6}$

29. $\dfrac{1}{2\sqrt{x+1}}$ **31.** $-\dfrac{3}{2\sqrt{1-3x}}$

Exercises 22-4, page 664

1. $m = 4$

3. $m = -1$

5. 4.00, 4.00, 4.00, 4.00, 4.00; $\lim\limits_{t\to 3} v = 4$ m/s

7. 5, 6.5, 7.7, 7.97, 7.997; $\lim\limits_{t\to 2} v = 8$ m/s

9. 4; 4 m/s   **11.** $6t - 4$; 8 m/s

13. $3 + \dfrac{2}{t^2}$ **15.** $6t - 6t^2$

17. $12t - 4$ **19.** $6t$ **21.** -2 **23.** $6w$ **25.** 460 W   **27.** -83.1 W/m²·h   **29.** πd^2 **31.** $24.2/\sqrt{\lambda}$

Exercises 22-5, page 669

1. $5x^4$ **3.** $-36x^8$ **5.** $4x^3$ **7.** $2x + 2$ **9.** $15x^2 - 1$ **11.** $8x^7 - 28x^6 - 1$ **13.** $-42x^6 + 15x^2$

15. $x^2 + x$ **17.** 16   **19.** 33   **21.** 360   **23.** 14   **25.** $30t^4 - 5$ **27.** $-6 - 6t^2$ **29.** 64

31. 45   **33.** 1   **35.** $3\pi r^2$ **37.** 84 W/A   **39.** $a(c_1 + 2c_2 E + 3c_3 E^2)$ **41.** -12 N/cm   **43.** 391 mm²

Exercises 22-6, page 674

1. $x^2(3) + (3x + 2)(2x) = 9x^2 + 4x$ **3.** $6x(6x - 5) + (3x^2 - 5x)(6) = 54x^2 - 60x$

5. $(x + 2)(2) + (2x - 5)(1) = 4x - 1$

7. $(x^4 - 3x^2 + 3)(-6x^2) + (1 - 2x^3)(4x^3 - 6x) = -14x^6 + 30x^4 + 4x^3 - 18x^2 - 6x$

9. $(2x - 7)(-2) + (5 - 2x)(2) = -8x + 24$

11. $(x^3 - 1)(4x - 1) + (2x^2 - x - 1)(3x^2) = 10x^4 - 4x^3 - 3x^2 - 4x + 1$

13. $\dfrac{3}{(2x+3)^2}$ **15.** $\dfrac{-2x}{(x^2+1)^2}$ **17.** $\dfrac{6x - 2x^2}{(3-2x)^2}$ **19.** $\dfrac{-6x^2 + 6x + 4}{(3x^2+2)^2}$ **21.** $\dfrac{-x^2 - 16x - 6}{(x^2+x+2)^2}$

23. $\dfrac{-2x^4 + 2x^3 + 5x^2 + 4x}{(x^3+2x^2)^2}$ **25.** -107 **27.** 19   **29.** $\dfrac{-12x^3 + 45x^2 - 14x}{(3x-7)^2}$ **31.** 12   **33.** 1, -1

35. $8t^3 - 45t^2 - 14t - 8$ **37.** -0.07 V/Ω **39.** 1.2°C/h   **41.** $\dfrac{2R(R+2r)}{3(R+r)^2}$ **43.** $\dfrac{E^2(R-r)}{(R+r)^3}$

Exercises 22-7, page 680

1. $\dfrac{1}{2x^{1/2}}$ **3.** $-\dfrac{2}{x^3}$ **5.** $-\dfrac{1}{x^{4/3}}$ **7.** $\dfrac{3}{2}x^{1/2} + \dfrac{1}{x^2}$ **9.** $10x(x^2 + 1)^4$ **11.** $-192x^2(7 - 4x^3)^7$

13. $\dfrac{2x^2}{(2x^3 - 3)^{2/3}}$ **15.** $\dfrac{8x}{(1 - x^2)^5}$ **17.** $\dfrac{24x^3}{(2x^4 - 5)^{1/4}}$ **19.** $\dfrac{-4x}{(1 - 8x^2)^{3/4}}$ **21.** $\dfrac{12x + 5}{(8x + 5)^{1/2}}$

23. $\dfrac{-16x - 22}{(4x + 3)^{1/2}(8x + 1)^2}$ **25.** $\dfrac{3}{10}$ **27.** $\dfrac{5}{36}$ **29.** $\dfrac{x^3(0) - 1(3x^2)}{x^6} = -3x^{-4}$ **31.** $x = 0$ **33.** 1

35. -1.35 cm/s   **37.** $\dfrac{-4.5 \times 10^5}{V^{5/2}}$, -4.50 kPa/cm³   **39.** -45.2 W/m²·h   **41.** $\dfrac{8a^3}{(4a^2 - \lambda^2)^{3/2}}$

43. $\dfrac{2(w + 1)}{(2w^2 + 4w + 4)^{1/2}}$

Exercises 22-8, page 684

1. $-\dfrac{3}{2}$ **3.** $\dfrac{6x + 1}{4}$ **5.** $\dfrac{x}{y}$ **7.** $\dfrac{2x}{5y^4}$ **9.** $\dfrac{2x}{2y + 1}$ **11.** $\dfrac{-3y}{3x + 1}$ **13.** $\dfrac{-2x - y^3}{3xy^2 + 3}$

15. $\dfrac{3(y^2 + 1)(y^2 - 2x + 1)}{(y^2 + 1)^2 - 6x^2y}$ **17.** $\dfrac{4(2y - x)^3 - 2x}{8(2y - x)^3 - 1}$ **19.** $\dfrac{-3x(x^2 + 1)^2}{y(y^2 + 1)}$ **21.** 3   **23.** $-\dfrac{108}{157}$ **25.** 1

27. $-\dfrac{x}{y}$ **29.** $\dfrac{r - R + 1}{r + 1}$ **31.** $\dfrac{2C^2r(12CSr - 20Cr - 3L)}{3(C^2r^2 - L^2)}$

Exercises 22-9, page 688

1. $y' = 3x^2 + 2x,\ y'' = 6x + 2,\ y''' = 6,\ y^{(n)} = 0\ (n \ge 4)$
3. $f'(x) = 3x^2 - 24x^3,\ f''(x) = 6x - 72x^2,\ f'''(x) = 6 - 144x,\ f^{(4)}(x) = -144,\ f^{(n)}(x) = 0\ (n \ge 5)$
5. $y' = -8(1 - 2x)^3,\ y'' = 48(1 - 2x)^2,\ y''' = -192(1 - 2x),\ y^{(4)} = 384,\ y^{(n)} = 0\ (n \ge 5)$
7. $f'(x) = (8x + 1)(2x + 1)^2,\ f''(x) = 12(2x + 1)(4x + 1),\ f'''(x) = 24(8x + 3),\ f^{(4)}(x) = 192,\ f^{(n)}(x) = 0\ (n \ge 5)$
9. $84x^5 - 30x^4$ **11.** $-\dfrac{1}{4x^{3/2}}$ **13.** $-\dfrac{12}{(8x - 3)^{7/4}}$ **15.** $\dfrac{12}{(1 - 2x)^{5/2}}$ **17.** $600(2 - 5x)^2$
19. $30(27x^2 - 1)(3x^2 - 1)^3$ **21.** $\dfrac{4}{(1 - x)^3}$ **23.** $\dfrac{2}{(x + 1)^3}$ **25.** $-\dfrac{9}{y^3}$ **27.** $-\dfrac{6(x^2 - xy + y^2)}{(2y - x)^3}$ **29.** $\dfrac{9}{125}$
31. $-\dfrac{13}{384}$ **33.** $-9.8\ \text{m/s}^2$ **35.** $-\dfrac{1.60}{(2t + 1)^{3/2}}$

Review Exercises for Chapter 22, page 689

1. -4 **3.** $\dfrac{1}{4}$ **5.** 1 **7.** $\dfrac{7}{3}$ **9.** $\dfrac{2}{3}$ **11.** -2 **13.** 5 **15.** $-4x$ **17.** $-\dfrac{4}{x^3}$ **19.** $\dfrac{1}{2\sqrt{x + 5}}$

21. $14x^6 - 6x$ **23.** $\dfrac{2}{x^{1/2}} + \dfrac{3}{x^2}$ **25.** $\dfrac{1}{(1 - x)^2}$ **27.** $-12(2 - 3x)^3$ **29.** $\dfrac{9x}{(5 - 2x^2)^{7/4}}$ **31.** $\dfrac{-15x^2 + 2x}{(1 - 6x)^{1/2}}$
33. $\dfrac{-2x - 3}{2x^2(4x + 3)^{1/2}}$ **35.** $\dfrac{2x - 6(2x - 3y)^2}{1 - 9(2x - 3y)^2}$ **37.** $\dfrac{5}{48}$ **39.** $\dfrac{74}{5}$ **41.** $36x^2 - \dfrac{2}{x^3}$ **43.** $\dfrac{56}{(1 + 4x)^3}$ **45.** f_2
47. -31 **49.** $-k + k^2t - \dfrac{1}{2}k^3t^2$ **51.** $-\dfrac{2k}{r^3}$ **53.** $0.4(0.01t + 1)^2(0.04t + 1)$ **55.** $\dfrac{2R(R^3 - 3r^2R + 2r^3)}{3(R^2 - r^2)^2}$
57. $-\dfrac{1}{4\pi\sqrt{C}(L + 2)^{3/2}}$ **59.** $\dfrac{40V_2^{0.4}}{V_1^{1.4}}$ **61.** $0.049/\text{m}$ **63.** $p = 2w + \dfrac{150}{w},\ \dfrac{dp}{dw} = 2 - \dfrac{150}{w^2}$
65. $A = 4x - x^3,\ \dfrac{dA}{dx} = 4 - 3x^2$
67. At $t = 5$ years, $dV/dt = -\$7500/\text{year}$ (rate of appreciation is decreasing)
$\qquad\qquad d^2V/dt^2 = \$1500/\text{year}^2$ (rate at which appreciation changes is increasing)
$\qquad$ (Machinery is depreciating, but depreciation is lessening.)

Exercises 23-1, page 696

1. $4x - y - 2 = 0$ **3.** $2y + x - 2 = 0$ **5.** $x - 2y + 6 = 0$ **7.** $8x - 4y - 7 = 0$

9. $\sqrt{3}x + 8y - 7 = 0,\ 16x - 2\sqrt{3}y - 15\sqrt{3} = 0$ **11.** $2x - 12y + 37 = 0,\ 72x + 12y + 37 = 0$ **13.** $y = 2x - 4$
15. $y - 8 = -\tfrac{1}{24}(x - \tfrac{3}{2})$, or $2x + 48y - 387 = 0$ **17.** $(-\tfrac{1}{4}, 0)$ **19.** $x - 2y - 20 = 0$ **21.** $x + y - 6 = 0$
23. $x + 2y - 3 = 0,\ x = 0,\ x - 2y + 3 = 0$

Exercises 23-2, page 700

1. $3.449\ 489\ 7$ **3.** $-0.180\ 460\ 4$ **5.** $0.585\ 786\ 4$ **7.** $0.348\ 894\ 2$ **9.** $2.561\ 552\ 8$ **11.** $-1.236\ 068\ 0$
13. $0.917\ 543\ 3$ **15.** $0.618\ 034\ 0$ **17.** $-1.855\ 772\ 5,\ 0.678\ 362\ 8,\ 3.177\ 409\ 7$ **19.** $1.587\ 401\ 1$ **21.** $29.4\ \text{m}$
23. $1.61\ \text{m}$

Exercises 23-3, page 705

1. 3.16, 341.6° **3.** 8.07, 352.4° **5.** $a = 0$ **7.** 20.0, 3.7°
9. 9.4 m/s, 302° **11.** 1.3 m/min², 288°
13. 41 m/s, 348°; 9.8 m/s², 270°
15. 276 m/s, 43.5°; 2090 m/s, 16.7°
17. 22.1 m/s², 25.4°; 20.2 m/s², 8.5° **19.** 21.2 km/min, 296.6°
21. $x^2 + y^2 = 4.445^2$; $v_x = 731$ m/min, $v_y = -690$ m/min
23. 371 m/s, 19.3°

Exercises 23-4, page 709

1. 0.0900 Ω/s **3.** \$1.22/week **5.** 4.1×10^{-6} m/s **7.** $\dfrac{dB}{dt} = \dfrac{-3kr(dr/dt)}{[r^2 + (\ell/2)^2]^{5/2}}$ **9.** 0.38 mm²/month
11. −101 mm³/min **13.** −3.75 kPa/min **15.** 2.51×10^6 mm³/s **17.** 0.48 m/min **19.** 2.71 m/s
21. 825 km/h **23.** 2.50 m/s

Exercises 23-5, page 716

1. Inc. $x > -1$, dec. $x < -1$ **3.** Inc. $-2 < x < 2$, dec. $x < -2$, $x > 2$ **5.** Min. $(-1, -1)$
7. Min. $(-2, -16)$, max. $(2, 16)$ **9.** Conc. up all x **11.** Conc. up $x < 0$, conc. down $x > 0$, infl. $(0, 0)$
13.   **15.**   **17.** Max. $(3, 18)$,   **19.** Max. $(-2, 8)$, min. $(0, 0)$,
 conc. down all x infl. $(-1, 4)$

21. No max. or min.,   **23.** Max. $(-1, 4)$, min. $(1, -4)$,   **25.** Max. $(1, 1)$, infl. $(0, 0)$, $(\frac{2}{3}, \frac{16}{27})$
 infl. $(-1, 1)$ infl. $(0, 0)$

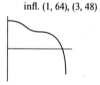

27. Max. $(2000, 1000)$ **29.** Max. $(0, 75)$,   **31.** $V = 4x^3 - 40x^2 + 96x$,
    infl. $(1, 64)$, $(3, 48)$ max. $(1.57, 67.6)$

33.   **35.**

110

Exercises 23-6, page 720

1. Inc. $x < 0$, dec. $x > 0$,
conc. up $x < 0$, $x > 0$,
asym. $x = 0$, $y = 0$

3. Dec. $x < -1$, $x > -1$,
conc. up $x > -1$,
conc. down $x < -1$,
int. $(0, 2)$, asym. $x = -1$, $y = 0$

5. Int. $(-\sqrt[3]{2}, 0)$, min. $(1, 3)$,
infl. $(-\sqrt[3]{2}, 0)$, asym. $x = 0$

7. Int. $(1, 0)$, $(-1, 0)$,
asym. $x = 0$, $y = x$,
conc. up $x < 0$, conc. down $x > 0$

9. Int. $(0, 0)$, max. $(-2, -4)$
min. $(0, 0)$, asym. $x = -1$

11. Int. $(0, -1)$, max. $(0, -1)$,
asym. $x = 1$, $x = -1$, $y = 0$

13. Int. $(1, 0)$, max. $(2, 1)$
infl. $(3, \frac{8}{9})$,
asym. $x = 0$, $y = 0$

15. Int. $(0, 0)$, $(1, 0)$, $(-1, 0)$
max. $(\frac{1}{2}\sqrt{2}, \frac{1}{2})$,
min. $(-\frac{1}{2}\sqrt{2}, -\frac{1}{2})$

17. Int. $(0, 0)$, infl. $(0, 0)$,
asym. $x = -3$, $x = 3$, $y = 0$

19. Int. $(0, 0)$, asym. $C_T = 6$,
inc. $C \geq 0$,
conc. down $C \geq 0$

21. Int. $(0, 1)$, max. $(0, 1)$,
infl. $(141, 0.82)$,
asym. $R = 0$

23. $A = 2\pi r^2 + \dfrac{40}{r}$,
min. $(1.47, 40.8)$,
asym. $r = 0$

Exercises 23-7, page 726

1. 60.0 m   **3.** $\dfrac{E}{2R}$ **5.** 34.6 m², \$8310   **7.** 0.250 cm/s   **9.** 8, 8   **11.** 5 mm, 5 mm   **13.** 17.0 cm

15. 8.49 cm, 8.49 cm   **17.** 1.20 m   **19.** 12   **21.** $0.58L$ **23.** 3.33 cm   **25.** 100 m
27. $w = 0.500$ m, $d = 0.866$ m   **29.** 8.00 km from refinery   **31.** 59.2 m, 118 m

Review Exercises for Chapter 23, page 728

1. $5x - y + 1 = 0$ **3.** $27x - 3y - 26 = 0$ **5.** $x - 2y + 3 = 0$ **7.** 4.19, 72.6°   **9.** 2.12   **11.** 2.00, 90.9°
13. 0.745 898 3   **15.** 1.911 164 3

17. Min. $(-2, -16)$,
conc. up all x

19. Int. $(0, 0)$, $(\pm 3\sqrt{3}, 0)$,
max. $(3, 54)$, min. $(-3, -54)$,
infl. $(0, 0)$

21. Min. $(2, -48)$,
conc. up $x < 0$, $x > 0$

111

23. Int. (0, 0), asym. $x = -1$, $y = 1$

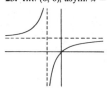

43. Int. (1, 0),
max. $(2, \frac{1}{4})$,
infl. $(3, \frac{2}{9})$,
asym. $P = 0$,
$V \geq 1$ (only meaningful values)

25. $2x - y + 1 = 0$
29. 8.8 m/s, 336°
33. $2400
35.

45. 1160 km/h

27. 0.0 m, 6.527 m
31. -7.44 cm/s

37. 1.30 cm/min
39. 37 700 m²/min
41. 5000 cm²

47. 6.6 cm

Exercises 24-1, page 735

1. $(5x^4 + 1)\,dx$ **3.** $\dfrac{-10\,dx}{x^6}$ **5.** $8x(x^2 - 1)^3\,dx$ **7.** $\dfrac{-12x\,dx}{(3x^2 + 1)^2}$ **9.** $x(1 - x)^2(-5x + 2)\,dx$ **11.** $\dfrac{2\,dx}{(5x + 2)^2}$

13. 12.28, 12   **15.** 1.712 75, 1.675   **17.** -2.4, $-2.473\,088\,1$ **19.** 0.6257, 0.626 490 3

21. 0.0038 cm²   **23.** -8.3 nF   **25.** 3.1 m³   **27.** $\dfrac{dr}{r} = \dfrac{k\,d\lambda/2\lambda^{1/2}}{k\lambda^{1/2}}$ **29.** $\dfrac{dA}{A} = \dfrac{2\,ds}{s}$ **31.** 16.96

Exercises 24-2, page 738

1. x^3 **3.** x^6 **5.** $\frac{3}{2}x^4 + x$ **7.** $\frac{2}{3}x^3 - \frac{1}{2}x^2$ **9.** $x^{5/2}$ **11.** $\frac{4}{3}x^{3/2} + 3x$ **13.** $\dfrac{1}{x}$ **15.** $\dfrac{2}{x^3}$ **17.** $\frac{2}{5}x^5 + x$

19. $3x^2 - \dfrac{1}{3x^3}$ **21.** $\frac{1}{3}x^3 + 2x - \dfrac{1}{x}$ **23.** $(2x + 1)^6$ **25.** $(x^2 - 1)^4$ **27.** $\frac{1}{40}(2x^4 + 1)^5$ **29.** $(6x + 1)^{3/2}$
31. $\frac{1}{4}(3x + 1)^{4/3}$

Exercises 24-3, page 744

1. $x^2 + C$ **3.** $\frac{1}{8}x^8 + C$ **5.** $\frac{2}{5}x^{5/2} + C$ **7.** $-\dfrac{1}{3x^3} + C$ **9.** $\frac{1}{3}x^3 - \frac{1}{6}x^6 + C$ **11.** $3x^3 + \frac{1}{2}x^2 + 3x + C$

13. $-\dfrac{1}{2x^2} + \dfrac{1}{2}x + C$ **15.** $\frac{2}{7}x^{7/2} - \frac{2}{5}x^{5/2} + C$ **17.** $6x^{1/3} + \frac{1}{9}x + C$ **19.** $\frac{1}{6}(1 + 2x)^3 + C$
21. $\frac{1}{6}(x^2 - 1)^6 + C$ **23.** $\frac{1}{5}(x^4 + 3)^5 + C$ **25.** $\frac{1}{40}(x^5 + 4)^8 + C$ **27.** $\frac{1}{12}(8x + 1)^{3/2} + C$
29. $\frac{1}{6}\sqrt{6x^2 + 1} + C$ **31.** $\sqrt{x^2 - 2x} + C$ **33.** $y = 2x^3 + 2$ **35.** $y = 5 - \frac{1}{18}(1 - x^3)^6$
37. $12y = 83 + (1 - 4x^2)^{3/2}$ **39.** $i = 2t^2 - 0.2t^3 + 2$ **41.** $f = \sqrt{0.01A + 1} - 1$ **43.** $y = 3x^2 + 2x - 3$

Exercises 24-4, page 750

1. 9, 12.15   **3.** 1.92, 2.28   **5.** 7.625, 8.208   **7.** 0.464, 0.5995   **9.** 13.5   **11.** $\frac{8}{3}$ **13.** 9   **15.** 0.8

Exercises 24-5, page 753

1. 1   **3.** $\frac{254}{7}$ **5.** $6 + 2\sqrt{6} - 2\sqrt{3}$ **7.** $\frac{3}{2}\sqrt[3]{2}$ **9.** $\frac{747}{20}$ **11.** $\frac{33}{20}\sqrt[3]{\frac{11}{5}} - \frac{3}{8}\sqrt[3]{\frac{1}{2}} - \frac{17}{5} = -1.552$ **13.** $\frac{4}{3}$
15. $\frac{81}{4}$ **17.** 2   **19.** $\frac{1}{4}(20.5^{2/3} - 17.5^{2/3}) = 0.1875$ **21.** $\frac{88}{3249} = 0.0271$ **23.** 49   **25.** $\frac{364}{3}$ **27.** $\frac{110}{3}$
29. $\frac{33}{784} = 0.0421$ **31.** $\frac{464}{5}$ **33.** 64 000 N·m   **35.** 86.8 m²

Exercises 24-6, page 757

1. $\frac{11}{2} = 5.50$, $\frac{16}{3} = 5.33$ **3.** 7.661, $\frac{23}{3} = 7.667$ **5.** 0.2042   **7.** 2.996   **9.** 0.5205   **11.** 21.74   **13.** 45.36
15. $0.177k$

Exercises 24-7, page 760

1. (a) 6, (b) 6   **3.** (a) 19.67, (b) 19.67   **5.** 0.2028   **7.** 3.084   **9.** 0.5114   **11.** 13.085   **13.** 44.63
15. 1.200 cm

Review Exercises for Chapter 24, page 761

1. $x^4 - \frac{1}{2}x^2 + C$ **3.** $x^2 - \frac{4}{5}x^{5/2} + C$ **5.** $\frac{20}{3}$ **7.** $\frac{16}{3}$ **9.** $3x - \frac{1}{x^2} + C$ **11.** $\sqrt{2}$ **13.** $\frac{1}{5(2-5x)} + C$

15. $-\frac{6}{7}(7-2x)^{7/4} + C$ **17.** $\frac{9}{8}(3\sqrt[3]{3}-1)$ **19.** $-\frac{1}{30}(1-2x^3)^5 + C$ **21.** $-\frac{1}{2x-x^3} + C$ **23.** $\frac{3350}{3}$

25. $\frac{-6x\,dx}{(x^2-1)^4}$ **27.** $\frac{(1-4x)\,dx}{(1-3x)^{2/3}}$ **29.** 0.061   **31.** $y = 3x - \frac{1}{3}x^3 + \frac{17}{3}$
33. (a) $x - x^2 + C_1$, (b) $-\frac{1}{4}(1-2x)^2 + C_2 = x - x^2 + C_2 - \frac{1}{4}$; $C_1 = C_2 - \frac{1}{4}$ **35.** 22   **37.** 0.842
39. 0.811   **41.** 1.01   **43.** 13.77   **45.** 1.85 m^3 **47.** $\frac{R\,dR}{R^2 + X^2}$ **49.** $y = k(2L^3x - 6Lx^2 + \frac{2}{5}x^5)$
51. 14.9 m^2

Exercises 25-1, page 769

1. 24.5 m/s   **3.** $s = 8.00 - 0.25t$ **5.** 5.0 m/s   **7.** 17 800 m   **9.** 24 m/s   **11.** 85.3 m   **13.** 0.345 nC
15. 0.017 C   **17.** 667 V   **19.** 4.65 mV   **21.** 0.55   **23.** 66.7 A   **25.** $\frac{k}{x_1}$
27. $m = 1002 - 2\sqrt{t+1}$, 2.51×10^5 min

Exercises 25-2, page 775

1. 2   **3.** $\frac{8}{3}$ **5.** $\frac{27}{8}$ **7.** $\frac{4}{3}\sqrt{2}$ **9.** $\frac{1}{2}$ **11.** $\frac{1}{6}$ **13.** $\frac{26}{3}$ **15.** 3   **17.** $\frac{15}{4}$ **19.** $\frac{7}{6}$ **21.** $\frac{256}{5}$ **23.** $\frac{7}{6}$
25. $\frac{81}{4}$ **27.** 1   **29.** 18.0 J   **31.** 80.8 km   **33.** 4 cm^2 **35.** 0.683 m^2

Exercises 25-3, page 782

1. $\frac{1}{3}\pi$ **3.** $\frac{1}{3}\pi$ **5.** 72π **7.** $\frac{768}{7}\pi$ **9.** $\frac{348}{5}\pi$ **11.** $\frac{16}{3}\pi$ **13.** $\frac{1}{3}\pi$ **15.** $\frac{1}{3}\pi$ **17.** $\frac{2}{5}\pi$ **19.** $\frac{8}{3}\pi$
21. $\frac{16}{15}\pi$ **23.** $\frac{16}{3}\pi$ **25.** $\frac{8}{3}\pi$ **27.** $\frac{1}{3}\pi r^2 h$ **29.** 7.56 mm^3 **31.** 18.3 cm^3

Exercises 25-4, page 789

1. $(\frac{10}{3}, 0)$ **3.** $(\frac{14}{15}, 0)$ **5.** $(-\frac{1}{2}, \frac{1}{2})$ **7.** $(\frac{7}{22}, \frac{5}{22})$ **9.** $(0, \frac{6}{5})$ **11.** $(\frac{4}{3}, \frac{4}{3})$ **13.** $(\frac{3}{5}, \frac{12}{35})$ **15.** $(0, \frac{5}{6})$
17. $(\frac{2}{3}, 0)$ **19.** $(\frac{2}{3}a, \frac{1}{3}b)$ **21.** 0.375 cm above center of base   **23.** 19.6 cm from larger base

Exercises 25-5, page 794

1. 128, 4   **3.** 214, 3.27   **5.** $\frac{64}{15}k$ **7.** $\frac{2}{3}\sqrt{6}$ **9.** $\frac{1}{6}ma^2$ **11.** $\frac{4}{7}\sqrt{7}$ **13.** $\frac{8}{11}\sqrt{55}$ **15.** $\frac{64}{3}\pi k$
17. $\frac{2}{5}\sqrt{10}$ **19.** $\frac{3}{10}mr^2$ **21.** 0.324 g·cm^2 **23.** 31.2 kg·cm^2

Exercises 25-6, page 799

1. 8.0 N·cm   **3.** 200 N·mm   **5.** 9.4×10^{-22} N·m   **7.** 1800 N·m   **9.** 3.00×10^6 kN·m
11. 9.85×10^5 N·m   **13.** 12.5 kN   **15.** 152 kN   **17.** 6530 N   **19.** 1.84 MN   **21.** 2.7 A   **23.** 35.3%
25. 109 m   **27.** $A = \pi r\sqrt{r^2 + h^2}$

Review Exercises for Chapter 25, page 802

1. 4.3 s   **3.** 4.7 s   **5.** 0.44 C   **7.** 3640 V   **9.** $y = 20x + \frac{1}{120}x^3$ **11.** $\frac{2}{3}$ **13.** 18   **15.** $\frac{27}{4}$ **17.** $\frac{48}{5}\pi$
19. $\frac{243}{10}\pi$ **21.** $\frac{4}{3}\pi ab^2$ **23.** $(\frac{40}{21}, \frac{10}{3})$ **25.** $(\frac{14}{5}, 0)$ **27.** $\frac{8}{5}k$ **29.** $\frac{256}{3}\pi k$ **31.** 2700 N·m   **33.** 1.8 m
35. 47 m^3 **37.** 1580 kN   **39.** 0.29 Ω

Exercises 26-1, page 809

1. $\cos(x + 2)$ **3.** $4\cos(2x - 1)$ **5.** $3\sin\frac{1}{2}x$ **7.** $-6\sin(3x - 1)$ **9.** $8\sin 4x \cos 4x = 4\sin 8x$
11. $-45\cos^2(5x + 2)\sin(5x + 2)$ **13.** $\sin 3x + 3x\cos 3x$ **15.** $9x^2\cos 5x - 15x^3\sin 5x$
17. $2x\cos x^2\cos 2x - 2\sin x^2\sin 2x$ **19.** $\frac{2\cos 4x}{\sqrt{1 + \sin 4x}}$ **21.** $\frac{3x\cos 3x - \sin 3x}{x^2}$
23. $\frac{4x(1 - 3x)\sin x^2 - 6\cos x^2}{(3x - 1)^2}$ **25.** $4\sin 3x(3\cos 3x \cos 2x - \sin 3x \sin 2x)$

113

27. $\dfrac{-2 \cos 3x[3 \sin 3x(1 + 2 \sin^2 2x) + 4 \cos 3x \sin 2x \cos 2x]}{(1 + 2 \sin^2 2x)^2}$ **29.** $3 \sin^2 x \cos x + 2 \sin 2x$

31. $1 - 4 \sin^2 4x \cos 4x$ **33.** See the table.

35. (a) 0.540 302 3, value of derivative, (b) 0.540 260 2, slope of secant line   **37.** Resulting curve is $y = \cos x$.

39. $\dfrac{2x - y \cos xy}{x \cos xy - 2 \sin 2y}$ **41.** $\dfrac{d \sin x}{dx} = \cos x, \dfrac{d^2 \sin x}{dx^2} = -\sin x, \dfrac{d^3 \sin x}{dx^3} = -\cos x, \dfrac{d^4 \sin x}{dx^4} = \sin x$

43. $\sin 2x = 2 \sin x \cos x$ **45.** 0.085   **47.** -2.36 **49.** -199 cm/s   **51.** -38.5 km

Exercises 26-2, page 813

1. $5 \sec^2 5x$ **3.** $2(1 - x) \csc^2 (1 - x)^2$ **5.** $6 \sec 2x \tan 2x$ **7.** $\dfrac{3 \csc \sqrt{x} \cot \sqrt{x}}{2\sqrt{x}}$ **9.** $30 \tan 3x \sec^2 3x$

11. $-4 \cot^3 \tfrac{1}{2}x \csc^2 \tfrac{1}{2}x$ **13.** $2 \tan 4x \sqrt{\sec 4x}$ **15.** $-84 \csc^4 7x \cot 7x$ **17.** $x^2 \sec^2 x + 2x \tan x$

19. $-4 \csc x^2(2x \cos x \cot x^2 + \sin x)$ **21.** $-\dfrac{\csc x(x \cot x + 1)}{x^2}$

23. $\dfrac{-4 \sin 4x - 4 \sin 4x \cot 3x + 3 \cos 4x \csc^2 3x}{(1 + \cot 3x)^2}$ **25.** $\sec^2 x(\tan^2 x - 1)$ **27.** $2 \sec 2x(\sec 2x - \tan 2x)$

29. $\dfrac{1 + 2 \sec^2 4x}{\sqrt{2x + \tan 4x}}$ **31.** $\dfrac{2 \cos 2x - \sec y}{x \sec y \tan y - 2}$ **33.** $24 \tan 3x \sec^2 3x \, dx$ **35.** $4 \sec 4x(\tan^2 4x + \sec^2 4x) \, dx$

37. (a) 3.425 518 8, value of derivative, (b) 3.426 052 4, slope of secant line   **39.** $2 \tan x \sec^2 x = 2 \sec x(\sec x \tan x)$

41. -12 **43.** $2 \sec^2 x - \sec x \tan x = \dfrac{2}{\cos^2 x} - \dfrac{\sin x}{\cos^2 x}$ **45.** -8.4 cm/s   **47.** 140 m/s

Exercises 26-3, page 817

1. $\dfrac{2x}{\sqrt{1 - x^4}}$ **3.** $\dfrac{18x^2}{\sqrt{1 - 9x^6}}$ **5.** $-\dfrac{1}{\sqrt{4 - x^2}}$ **7.** $\dfrac{1}{\sqrt{(x - 1)(2 - x)}}$ **9.** $\dfrac{1}{2\sqrt{x}(1 + x)}$ **11.** $-\dfrac{1}{x^2 + 1}$

13. $\dfrac{x}{\sqrt{1 - x^2}} + \text{Arcsin } x$ **15.** $\dfrac{4x}{1 + 4x^2} + 2 \text{ Arctan } 2x$ **17.** $\dfrac{3\sqrt{1 - 4x^2} \text{ Arcsin } 2x - 6x + 2}{\sqrt{1 - 4x^2} \text{ Arcsin}^2 2x}$

19. $\dfrac{2(\text{Arccos } 2x + \text{Arcsin } 2x)}{\sqrt{1 - 4x^2} \text{ Arccos}^2 2x}$ **21.** $\dfrac{-24 \text{ Arccos}^2 4x}{\sqrt{1 - 16x^2}}$ **23.** $\dfrac{8 \text{ Arcsin } 4x}{\sqrt{1 - 16x^2}}$ **25.** $\dfrac{9 \text{ Arctan}^2 x}{1 + x^2}$ **27.** $\dfrac{-2(2x + 1)^2}{(1 + 4x^2)^2}$

29. $\dfrac{18(4 - \text{Arccos } 2x)^2}{\sqrt{1 - 4x^2}}$ **31.** $-\dfrac{x^2y^2 + 2y + 1}{2x}$

33. (a) 1.154 700 5, value of derivative, (b) 1.154 739 0, slope of secant line   **35.** $\dfrac{3 \text{ Arcsin}^2 x \, dx}{\sqrt{1 - x^2}}$ **37.** $\dfrac{-16x}{(1 + 4x^2)^2}$

39. Let $y = \text{Arcsec } u$; solve for u; take derivatives; substitute.   **41.** $\dfrac{E - A}{\omega m \sqrt{m^2 E^2 - (A - E)^2}}$

43. $\theta = \text{Arctan } \dfrac{h}{x}; \dfrac{d\theta}{dx} = \dfrac{-h}{x^2 + h^2}$

Exercises 26-4, page 821

1. $d \sin x/dx = \cos x$ and $d \cos x/dx = -\sin x$, and $\sin x = \cos x$ at points of intersection.

3. $\dfrac{1}{x^2 + 1}$ is always positive.   **5.** Dec. $x > 0, x < 0$,   **7.** $8\sqrt{2}x + 8y + 4\sqrt{2} - 5\pi\sqrt{2} = 0$
infl. (0, 0),
asym. $x = \pi/2, x = -\pi/2$ **9.** 1.933 753 8   **11.** 10
13. 0.58 m/s, -1.7 m/s^2
15. -0.072 N/s
17. 251 cm/s, 270°
19. 9480 cm/s^2, 0°

21. -0.085 rad/s   **23.** 8.08 m/s   **25.** 0.020   **27.** 0.19 m   **29.** 100 mm^2 **31.** 4.2 m

114

Exercises 26-5, page 826

1. $\dfrac{2\log e}{x}$ **3.** $\dfrac{6\log_5 e}{3x+1}$ **5.** $\dfrac{-3}{1-3x}$ **7.** $\dfrac{4\sec^2 2x}{\tan 2x}=4\sec 2x\csc 2x$ **9.** $\dfrac{1}{2x}$ **11.** $\dfrac{6(x+1)}{x^2+2x}$

13. $2(1+\ln x)$ **15.** $\dfrac{3(2x+1)\ln(2x+1)-6x}{(2x+1)[\ln(2x+1)]^2}$ **17.** $\dfrac{1}{x\ln x}$ **19.** $\dfrac{1}{x^2+x}$ **21.** $\dfrac{\cos\ln x}{x}$ **23.** $\dfrac{6\ln 2x}{x}$

25. $\dfrac{x\sec^2 x+\tan x}{-x\tan x}$ **27.** $\dfrac{x+4}{x(x+2)}$ **29.** $\dfrac{\sqrt{x^2+1}}{x}$ **31.** $\dfrac{x+y-2\ln(x+y)}{x+y+2\ln(x+y)}$

33. 0.5 is value of derivative; 0.499 987 5 is slope of secant line. **35.** See the table **37.** 2.73

39. $-2(\tan x+\sec x\csc x)\,dx$ **41.** -1 **43.** $x^x(\ln x+1)$ **45.** $\dfrac{10\log e}{I}\dfrac{dI}{dt}$ **47.** 0.125 s²/m

Exercises 26-6, page 829

1. $(2\ln 3)3^{2x}$ **3.** $(6\ln 4)4^{6x}$ **5.** $6e^{6x}$ **7.** $\dfrac{e^{\sqrt x}}{2\sqrt x}$ **9.** $e^{-x}(1-x)$ **11.** $e^{\sin x}(x\cos x+1)$

13. $\dfrac{3e^{2x}(2x+1)}{(x+1)^2}$ **15.** $e^{-3x}(4\cos 4x-3\sin 4x)$ **17.** $\dfrac{2e^{3x}(12x+5)}{(4x+3)^2}$ **19.** $\dfrac{2xe^{x^2}}{e^{x^2}+4}$

21. $16e^{6x}(x\cos x^2+3\sin x^2)$ **23.** $2(\ln 2x+e^{2x})\left(\dfrac{1}{x}+2e^{2x}\right)$ **25.** $\dfrac{e^{xy}(xy+1)}{1-x^2e^{xy}-\cos y}$ **27.** $\dfrac{e^{2x}}{x}+2e^{2x}\ln x$

29. $12e^{6x}\cot 2e^{6x}$ **31.** $\dfrac{4e^{2x}}{\sqrt{1-e^{4x}}}$ **33.** (a) 2.718 281 8, value of derivative, (b) 2.718 417 8, slope of secant line

35. -0.724 **37.** $(-xe^{-x}+e^{-x})+(xe^{-x})=e^{-x}$ **39.** $\dfrac{dy}{dx}=e^a;\ y=e^a$ **41.** $(2-t)e^{-0.5t}$ **43.** $-0.001\,64/\text{h}$

45. Substitute and simplify. **47.** $\dfrac{d}{dx}\left[\dfrac{1}{2}(e^u-e^{-u})\right]=\dfrac{1}{2}(e^u+e^{-u})\dfrac{du}{dx};\ \dfrac{d}{dx}\left[\dfrac{1}{2}(e^u+e^{-u})\right]=\dfrac{1}{2}(e^u-e^{-u})\dfrac{du}{dx}$

Exercises 26-7, page 833

1. Int. (0, 0), max. (0, 0), not defined for $\cos x<0$, asym. $x=-\tfrac12\pi,\ \tfrac12\pi,\ldots$

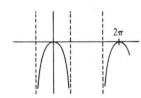

3. Int. (0, 0), max. $\left(1,\dfrac{1}{e}\right)$, infl. $\left(2,\dfrac{2}{e^2}\right)$, asym. $y=0$

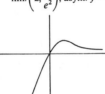

5. Int. (0, 0), max. (0, 0), infl. $(-1,-\ln 2),\ (1,-\ln 2)$

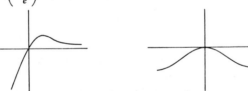

7. Int. (0, 1), max. (0, 1), infl. $(\tfrac12\sqrt2,\tfrac1e\sqrt e),\ (-\tfrac12\sqrt2,\tfrac1e\sqrt e)$, asym. $y=0$

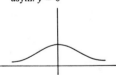

9. Max. $(1,-1)$, asym. $x=0$

11. Int. (0, 0), infl. (0, 0), inc. all x

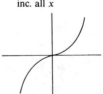

13. $y=x-1$ **15.** $2\sqrt2 x-2y+2\sqrt2-3\pi\sqrt2=0$ **17.** 1.314 096 8 **19.** -0.303 W/day

21. $\dfrac{p(-a+bT)}{T^2}$ **23.** $a=k^2x$ **25.** Int. (0, 0), min. (0, 0), (2π, 0), . . . , asym. $x=-\tfrac{\pi}{2},\tfrac{\pi}{2},\tfrac{3\pi}{2},\ldots$

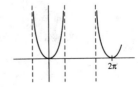

27. $dR_T=-\tfrac12ke^{k\,\csc(\theta/2)}\csc\tfrac\theta2\cot\tfrac\theta2\,d\theta$

29. $v=-e^{-0.5t}(1.4\cos 6t+2.3\sin 6t),\ -2.03$ cm/s **31.** $1/\sqrt e=0.607$

Review Exercises for Chapter 26, page 835

1. $-12 \sin(4x - 1)$ **3.** $-\dfrac{\sec^2 \sqrt{3-x}}{2\sqrt{3-x}}$ **5.** $-6 \csc^2(3x + 2) \cot(3x + 2)$ **7.** $-24x \cos^3 x^2 \sin x^2$

9. $2e^{2(x-3)}$ **11.** $\dfrac{6x}{x^2 + 1}$ **13.** $\dfrac{9}{9 + x^2}$ **15.** $\dfrac{4}{(\text{Arcsin } 4x)(\sqrt{1 - 16x^2})}$ **17.** $(-2 \csc 4x)\sqrt{\csc 4x + \cot 4x}$

19. $\dfrac{2(1 + e^{-x})}{x - e^{-x}}$ **21.** $\dfrac{-\cos x(2e^{3x} \sin x + 3e^{3x} \cos x + 2 \sin x)}{(e^{3x} + 1)^2}$ **23.** $\dfrac{2x(1 + 4x^2) \text{ Arctan } 2x - 2x^2}{(1 + 4x^2)(\text{Arctan } 2x)^2}$

25. $-2x \cot x^2$ **27.** $\dfrac{2 \cos x \ln(3 + \sin x)}{3 + \sin x}$ **29.** $e^{-2x} \sec x(\tan x - 2)$ **31.** $\dfrac{\cos 2x + 2e^{4x}}{\sqrt{\sin 2x + e^{4x}}}$

33. $\dfrac{2x^3 e^y + 2xy^2 e^y + y}{x - x^4 e^y - x^2 y^2 e^y}$ **35.** $2x(e^{2\cos^2 x})(1 - 2x \sin x \cos x)$ **37.** $\dfrac{y(xye^{-x} - 1)}{x(ye^{-x} + 1)}$ **39.** Arccos x

41. Infl. $(\frac{1}{2}\pi, \frac{1}{2}\pi)$, $(\frac{3}{2}\pi, \frac{3}{2}\pi)$ **43.** Max. $(e^{-2}, 4e^{-2})$,   **45.** $7.27x + y - 8.44 = 0$
min. $(1, 0)$, infl. (e^{-1}, e^{-1})

47. $2x + 2.57y - 4.30 = 0$

49. $2 \sin x \cos x - 2 \cos x \sin x = 0$

51. $-0.703\,467\,4$

53. -5.17 cm/s

55. $-20 \sin 2t$ **57.** $\dfrac{ab}{x(b - x)}$

59. $-kE_0^2 \cos \frac{1}{2}\theta \sin \frac{1}{2}\theta$

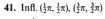

61. $\dfrac{f}{\sqrt{R^2 - F^2 f^2}}$ **63.** $0.005\,934$ rad/s   **65.** $-4.16°$C/min   **67.** 83.6 m/s   **69.** 2.00 m wide, 1.82 m high

71. $-0.006\,55$ rad/s   **73.** 0.40 cm/s, $72.5°$ **75.** 7.07 cm   **77.** $A = 100 \cos \theta(1 + \sin \theta)$ **79.** 10 items

Exercises 27-1, page 843

1. $\frac{1}{5} \sin^5 x + C$ **3.** $-\frac{2}{3}(\cos x)^{3/2} + C$ **5.** $\frac{4}{3} \tan^3 x + C$ **7.** $\frac{1}{8}$ **9.** $\frac{1}{4}(\text{Arcsin } x)^4 + C$

11. $\frac{1}{10}(\text{Arctan } 5x)^2 + C$ **13.** $\frac{1}{3}[\ln(x + 1)]^3 + C$ **15.** 0.179 **17.** $\frac{1}{4}(4 + e^x)^4 + C$ **19.** $\frac{3}{16}(2e^{2x} - 1)^{4/3} + C$

21. $\frac{1}{10}(1 + \sec^2 x)^5 + C$ **23.** $\frac{1}{6}$ **25.** 1.102 **27.** $y = \frac{1}{3}(\ln x)^3 + 2$ **29.** $\frac{1}{3}mnv^2$ **31.** $q = (1 - e^{-t})^3$

Exercises 27-2, page 846

1. $\frac{1}{4} \ln|1 + 4x| + C$ **3.** $-\frac{1}{3} \ln|4 - 3x^2| + C$ **5.** $\frac{1}{3} \ln 4 = 0.462$ **7.** $-\frac{1}{2} \ln|\cot 2x| + C$ **9.** $\ln 2 = 0.693$

11. $\ln|1 - e^{-x}| + C$ **13.** $\ln|x + e^x| + C$ **15.** $\frac{1}{4} \ln|1 + 4 \sec x| + C$ **17.** $\frac{1}{4} \ln 5 = 0.402$ **19.** $\ln|\ln x| + C$

21. $\ln|2x + \tan x| + C$ **23.** $-\sqrt{1 - 2x} + C$ **25.** $\ln|x| - \dfrac{2}{x} + C$ **27.** $\frac{1}{3} \ln(\frac{5}{4}) = 0.0744$ **29.** 1.10

31. $\pi \ln 2 = 2.18$ **33.** $y = \ln \dfrac{3.5}{3 + \cos x} + 2$ **35.** $P = P_0 e^{kt}$ **37.** $i = \dfrac{E}{R}(1 - e^{-Rt/L})$ **39.** 1.41 m

Exercises 27-3, page 850

1. $e^{7x} + C$ **3.** $\frac{1}{2}e^{2x+5} + C$ **5.** $2(e - 1) = 3.44$ **7.** $2e^{x^3} + C$ **9.** $2(e^2 - e) = 9.34$ **11.** $\frac{1}{2}e^{2 \sec x} + C$

13. $\dfrac{2e^x - 3}{2e^{2x}} + C$ **15.** $6 - \dfrac{3(e^6 - e^2)}{2} = -588.06$ **17.** $-\dfrac{4}{e^{\sqrt{x}}} + C$ **19.** $e^{\text{Arctan } x} + C$ **21.** $-\frac{1}{3}e^{\cos 3x} + C$

23. 0 **25.** 6.389 **27.** $\pi(e^4 - e) = 163$ **29.** $\frac{1}{8}(e^8 - 1) = 372$ **31.** $\ln b \int b^u \, du = b^u + C_1$

33. $q = EC(1 - e^{-t/RC})$ **35.** $s = -e^{-2t} - 0.6e^{-5t}$

Exercises 27-4, page 853

1. $\frac{1}{2} \sin 2x + C$ **3.** $\frac{1}{3} \tan 3x + C$ **5.** $2 \sec \frac{1}{2}x + C$ **7.** 0.6365 **9.** $\frac{3}{2} \ln|\sec x^2 + \tan x^2| + C$

11. $\cos\left(\dfrac{1}{x}\right) + C$ **13.** $\frac{1}{2}\sqrt{3}$ **15.** $\frac{1}{5} \sec 5x + C$ **17.** $\frac{1}{2} \ln|\sec 2x + \tan 2x| + C$

19. $\frac{1}{2}(\ln|\sin 2x| + \sin 2x) + C$ **21.** $\csc x - \cot x - \ln|\csc x - \cot x| + \ln|\sin x| + C$ **23.** $\frac{1}{9}\pi + \frac{1}{3} \ln 2 = 0.580$

25. 0.347 **27.** $\pi\sqrt{3} = 5.44$ **29.** $\theta = 0.10 \cos 2.5t$ **31.** 0.7726 m

1. $\frac{1}{3}\sin^3 x + C$ **3.** $-\frac{1}{2}\cos 2x + \frac{1}{6}\cos^3 2x + C$ **5.** $\frac{2}{3}\sin^3 x - \frac{2}{5}\sin^5 x + C$ **7.** $\frac{1}{120}(64 - 43\sqrt{2})$
9. $\frac{1}{2}x - \frac{1}{4}\sin 2x + C$ **11.** $\frac{1}{2}x + \frac{1}{12}\sin 6x + C$ **13.** $\frac{1}{2}\tan^2 x + \ln|\cos x| + C$ **15.** $\frac{3}{4}$

17. $\frac{1}{6}\tan^3 2x - \frac{1}{2}\tan 2x + x + C$ **19.** $\frac{1}{15}\sec^5 3x - \frac{1}{9}\sec^3 3x + C$ **21.** $x - \frac{1}{2}\cos 2x + C$
23. $\frac{1}{4}\cot^4 x - \frac{1}{3}\cot^3 x + \frac{1}{2}\cot^2 x - \cot x + C$ **25.** $1 + \frac{1}{2}\ln 2 = 1.347$ **27.** $\frac{1}{5}\tan^5 x + \frac{2}{3}\tan^3 x + \tan x + C$
29. $\frac{1}{2}\pi^2 = 4.935$ **31.** $\sqrt{2} - 1 = 0.414$ **33.** $\int \sin x \cos x \, dx = \frac{1}{2}\sin^2 x + C_1 = -\frac{1}{2}\cos^2 x + C_2; C_2 = C_1 + \frac{1}{2}$
35. $\frac{4}{3}$ **37.** 120 V   **39.** $\dfrac{aA}{2} + \dfrac{A}{2b\pi}\sin ab\pi \cos 2bc\pi$

1. $\text{Arcsin}\,\frac{1}{2}x + C$ **3.** $\frac{1}{8}\text{Arctan}\,\frac{1}{8}x + C$ **5.** $\frac{1}{4}\text{Arcsin}\,4x + C$ **7.** $\text{Arctan}\,6 = 1.41$
9. $\frac{2}{5}\sqrt{5}\,\text{Arcsin}\,\frac{1}{5}\sqrt{5} = 0.415$ **11.** $\frac{4}{9}\ln|9x^2 + 16| + C$ **13.** $\frac{1}{35}\sqrt{35}\,(\text{Arctan}\,\frac{2}{7}\sqrt{35} - \text{Arctan}\,\frac{1}{7}\sqrt{35}) = 0.057$
15. $\text{Arcsin}\,e^x + C$ **17.** $\text{Arctan}\,(x + 1) + C$ **19.** $4\,\text{Arcsin}\,\frac{1}{2}(x + 2) + C$ **21.** -0.357
23. $2\,\text{Arcsin}\,(\frac{1}{2}x) + \sqrt{4 - x^2} + C$ **25.** (a) Inverse tangent, (b) logarithmic, (c) general power

27. (a) General power, (b) inverse sine, (c) logarithmic   **29.** $\text{Arctan}\,2 = 1.11$ **31.** $k\,\text{Arctan}\,\dfrac{x}{d} + C$

33. $\text{Arcsin}\,\dfrac{x}{A} = \sqrt{\dfrac{k}{m}}\,t + \text{Arcsin}\,\dfrac{x_0}{A}$ **35.** $0.22k$

1. $\cos x + x \sin x + C$ **3.** $\frac{1}{2}xe^{2x} - \frac{1}{4}e^{2x} + C$ **5.** $x \tan x + \ln|\cos x| + C$
7. $2x\,\text{Arctan}\,x - 2\ln\sqrt{1 + x^2} + C$ **9.** $-8x\sqrt{1 - x} - \frac{16}{3}(1 - x)^{3/2} + C$ **11.** $\frac{1}{2}x^2 \ln x - \frac{1}{4}x^2 + C$
13. $\frac{1}{2}x \sin 2x - \frac{1}{4}(2x^2 - 1)\cos 2x + C$ **15.** $\frac{1}{2}(e^{\pi/2} - 1) = 1.91$ **17.** $1 - \dfrac{3}{e^2} = 0.594$ **19.** $\frac{1}{2}\pi - 1 = 0.571$
21. 0.756 **23.** $q = \frac{1}{5}[e^{-2t}(\sin t - 2\cos t) + 2]$

1. $-\dfrac{\sqrt{1 - x^2}}{x} - \text{Arcsin}\,x + C$ **3.** $2\ln|x + \sqrt{x^2 - 4}| + C$ **5.** $-\dfrac{\sqrt{x^2 + 9}}{9x} + C$ **7.** $\dfrac{x}{\sqrt{4 - x^2}} + C$

9. $\dfrac{16 - 9\sqrt{3}}{24} = 0.017$ **11.** $5\ln|\sqrt{x^2 + 2x + 2} + x + 1| + C$ **13.** $\frac{1}{3}\text{Arcsec}\,\frac{2}{3}x + C$ **15.** $\text{Arcsec}\,e^x + C$

17. π **19.** $\dfrac{1}{4}ma^2$ **21.** 2.68 **23.** $kQ \ln\dfrac{\sqrt{a^2 + b^2} + a}{\sqrt{a^2 + b^2} - a}$

1. $\frac{3}{25}[2 + 5x - 2\ln|2 + 5x|] + C$ **3.** $\frac{3544}{15} = 236.3$ **5.** $\frac{1}{2}x\sqrt{4 - x^2} + 2\,\text{Arcsin}\,\frac{1}{2}x + C$
7. $\frac{1}{2}\sin x - \frac{1}{10}\sin 5x + C$ **9.** $\sqrt{4x^2 - 9} - 3\,\text{Arcsec}\left(\dfrac{2x}{3}\right) + C$
11. $\frac{1}{20}\cos^4 4x \sin 4x + \frac{1}{5}\sin 4x - \frac{1}{15}\sin^3 4x + C$ **13.** $\frac{1}{2}x^2\,\text{Arctan}\,x^2 - \frac{1}{4}\ln(1 + x^4) + C$
15. $\frac{1}{4}(8\pi - 9\sqrt{3}) = 2.386$ **17.** $-\ln\left(\dfrac{1 + \sqrt{4x^2 + 1}}{2x}\right) + C$ **19.** $-\ln\left(\dfrac{1 + \sqrt{1 - 4x^2}}{2x}\right) + C$
21. $\frac{1}{8}\cos 4x - \frac{1}{12}\cos 6x + C$ **23.** $\frac{1}{3}(\cos x^3 + x^3 \sin x^3) + C$ **25.** $\dfrac{x^2}{\sqrt{1 - x^4}} + C$ **27.** 4.892

29. $\frac{1}{4}x^4(\ln x^2 - \frac{1}{2}) + C$ **31.** $-\dfrac{x^3}{3\sqrt{x^6 - 1}} + C$ **33.** $\frac{1}{4}[2\sqrt{5} + \ln(2 + \sqrt{5})] = 1.479$
35. $\frac{1}{4}(8\,\text{Arctan}\,4 - \ln 17) = 1.943$ **37.** 32.7 kN   **39.** 38.5 N·m

Review Exercises for Chapter 27, page 874

1. $-\frac{1}{2}e^{-2x} + C$ 3. $-\dfrac{1}{\ln 2x} + C$ 5. $4 \ln (1 + \sin x) + C$ 7. $\frac{2}{35}$ Arctan $\frac{7}{5}x + C$ 9. 0

11. $\frac{1}{2} \ln 2 = 0.3466$ 13. $\frac{1}{12} \sec^4 3x + C$ 15. $-\frac{1}{3} \ln |\cos 3x| + C$ 17. $\frac{1}{9} \tan^3 3x + \frac{1}{3} \tan 3x + C$

19. $\dfrac{3}{\sqrt{e}} - 2 = -0.1804$ 21. $\frac{3}{4}$ Arctan $\dfrac{x^2}{2} + C$ 23. $\frac{1}{2} \ln |2x + \sqrt{4x^2 - 9}| + C$ 25. $\sqrt{e^{2x} + 1} + C$

27. $\frac{1}{2}(3x - \sin 3x \cos 3x) + C$ 29. $-\frac{1}{2} \cot 2x + \frac{1}{4} \ln |\sin 2x| + C$ 31. $\frac{1}{2} \sin e^{2x} + C$ 33. $\frac{1}{3}$

35. $\frac{1}{2}x^2 - 2x + 3 \ln |x + 2| + C$ 37. $\frac{1}{3}(e^x + 1)^3 + C_1 = \frac{1}{3}e^{3x} + e^{2x} + e^x + C_2; C_2 = C_1 + \frac{1}{3}$

39. (a) $\dfrac{1}{2} \int (1 - \cos 2x) \, dx = \dfrac{x}{2} - \dfrac{1}{4} \sin 2x + C_1$,

(b) $\int \sin x(\sin x \, dx) = -\sin x \cos x + \int \cos^2 x \, dx = -\sin x \cos x + \int (1 - \sin^2 x) \, dx$ 41. $y = \frac{1}{3} \tan^3 x + \tan x$

43. $2(e^3 - 1) = 38.17$ 45. 11.18   47. $4\pi(e^2 - 1) = 80.29$ 49. $\frac{1}{8}\pi(e^{2\pi} - 1) = 209.9$ 51. $\ln 3$

53. $\Delta S = a \ln T + bT + \frac{1}{2}cT^2 + C$ 55. $v = 98(1 - e^{-t/10})$ 57. $\sqrt{2}$ 59. $\dfrac{2}{3} k$

61. 3.47 cm^3 63. 73.0 m^2

Exercises 28-1, page 880

1. 1, 4, 9, 16   3. $\frac{1}{2}, \frac{1}{3}, \frac{1}{4}, \frac{1}{5}$ 5. (a) $-\frac{2}{5}, \frac{4}{25}, -\frac{8}{125}, \frac{16}{625}$ (b) $-\frac{2}{5} + \frac{4}{25} - \frac{8}{125} + \frac{16}{625} - \cdots$

7. (a) 2, 0, 2, 0   (b) $2 + 0 + 2 + 0 + \cdots$ 9. $a_n = \dfrac{1}{n + 1}$ 11. $a_n = \dfrac{1}{(n + 1)(n + 2)}$

13. 1, 1.125, 1.162 037 0, 1.177 662 0, 1.185 662 0; convergent; 1.2

15. 1, 1.5, 2.166 666 7, 2.916 666 7, 3.716 666 7; divergent   17. 0, -1, -3, -6, -10; divergent

19. 0.75, 0.888 888 9, 0.937 500 0, 0.960 000 0, 0.972 222 2; convergent; 1   21. Divergent

23. Convergent, $S = \frac{3}{4}$ 25. Convergent, $S = 100$ 27. Convergent, $S = \frac{4096}{9}$ 29. (a) 1, (b) diverges

31. $r = x; S = \dfrac{x^0}{1 - x} = \dfrac{1}{1 - x}$

Exercises 28-2, page 885

1. $1 + x + \frac{1}{2}x^2 + \cdots$ 3. $1 - \frac{1}{2}x^2 + \frac{1}{24}x^4 - \cdots$ 5. $1 + \frac{1}{2}x - \frac{1}{8}x^2 + \cdots$ 7. $1 - 2x + 2x^2 - \cdots$

9. $1 - 8x^2 + \frac{32}{3}x^4 - \cdots$ 11. $1 + x + x^2 + \cdots$ 13. $-2x - 2x^2 - \frac{8}{3}x^3 - \cdots$ 15. $1 - \dfrac{x^2}{8} + \dfrac{x^4}{384} - \cdots$

17. $x - \frac{1}{3}x^3 + \cdots$ 19. $x + \frac{1}{3}x^3 + \cdots$ 21. $-\frac{1}{2}x^2 - \frac{1}{12}x^4 - \cdots$ 23. $x^2 - \frac{1}{3}x^4 + \cdots$

25. No, functions are not defined at $x = 0$.   27. $e^x = 1 + x + \dfrac{x^2}{2} + \cdots, e^{x^2} = 1 + x^2 + \dfrac{x^4}{2} + \cdots$

29. $f''(0) = 0$ except $f'''(0) = 6$ 31. $R = e^{-0.001t} = 1 - 0.001t + (5 \times 10^{-7})t^2 - \cdots$

Exercises 28-3, page 891

1. $1 + 3x + \frac{9}{2}x^2 + \frac{9}{2}x^3 + \cdots$ 3. $\dfrac{x}{2} - \dfrac{x^3}{2^3 3!} + \dfrac{x^5}{2^5 5!} - \dfrac{x^7}{2^7 7!} + \cdots$ 5. $1 - 8x^2 + \frac{32}{3}x^4 - \frac{256}{45}x^6 + \cdots$

7. $x^2 - \frac{1}{2}x^4 + \frac{1}{3}x^6 - \frac{1}{4}x^8 + \cdots$ 9. 0.3103   11. 0.1901   13. $1 + \frac{1}{2}x^2 + \frac{1}{24}x^4 + \frac{1}{720}x^6 + \cdots$

15. $x + x^2 + \frac{1}{3}x^3 + \cdots$ 17. $\dfrac{d}{dx}\left(x - \dfrac{1}{6}x^3 + \dfrac{1}{120}x^5 - \cdots\right) = 1 - \dfrac{1}{2}x^2 + \dfrac{1}{24}x^4 - \cdots$

19. $\int \cos x \, dx = x - \dfrac{x^3}{3!} + \cdots$ 21. $\int_0^1 e^x \, dx = 1.718 \, 281 \, 8, \int_0^1 \left(1 + x + \dfrac{1}{2}x^2 + \dfrac{1}{6}x^3\right) dx = 1.708 \, 333 \, 3$

23. 0.003 099   25. 0.1249   27. $y = -1 + 3x + 2t^2 - \dfrac{9}{2}t^3 - \cdots$

Exercises 28-4, page 894

1. 1.22, 1.221 402 8   3. 0.099 833 3, 0.099 833 4   5. 2.718 055 6, 2.718 281 8   7. 0.998 629 2, 0.998 629 5

9. 0.334 933 3, 0.336 472 2   11. 0.354 613 0, 0.354 612 9   13. $-0.013 \, 997 \, 5$, $-0.013 \, 997 \, 5$

15. 0.728 302 0, 0.728 695 0   17. 1.052 352 8   19. 0.987 446 2   21. 8.3×10^{-8} 23. 3.1×10^{-7}

25. 3.146   27. 1.59 years   29. $i = \dfrac{E}{L}\left(t - \dfrac{Rt^2}{2L}\right)$; small values of t 31. 7.8 m

Exercises 28-5, page 898

1. 3.32   **3.** 2.049   **5.** 0.5150   **7.** 0.492 88   **9.** $e^{-2}\left[1 - (x-2) + \dfrac{(x-2)^2}{2!} - \cdots\right]$

11. $\dfrac{1}{2}\left[\sqrt{3} + \left(x - \dfrac{1}{3}\pi\right) - \dfrac{\sqrt{3}}{2!}\left(x - \dfrac{1}{3}\pi\right)^2 - \cdots\right]$ **13.** $2 + \frac{1}{12}(x-8) - \frac{1}{288}(x-8)^2 + \cdots$

15. $1 + 2(x - \frac{1}{4}\pi) + 2(x - \frac{1}{4}\pi)^2 + \cdots$ **17.** 0.111   **19.** 3.0496   **21.** 2.0247   **23.** 0.874 62

25. Use the indicated method.   **27.** 0.515 040 8, 0.515 038 8, 0.515 038 1

Exercises 28-6, page 905

1. $f(x) = \frac{1}{2} - \frac{2}{\pi}\sin x - \frac{2}{3\pi}\sin 3x - \cdots.$

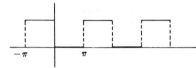

3. $f(x) = \frac{3}{2} + \frac{2}{\pi}\sin x + \frac{2}{3\pi}\sin 3x + \cdots$

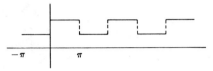

5. $f(x) = \frac{\pi}{4} - \frac{2}{\pi}(\cos x + \frac{1}{9}\cos 3x + \cdots) + (\sin x - \frac{1}{2}\sin 2x + \cdots)$

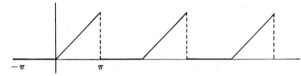

7. $f(x) = -\frac{1}{4} - \frac{1}{\pi}\cos x + \frac{1}{3\pi}\cos 3x - \cdots + \frac{3}{\pi}\sin x - \frac{1}{\pi}\sin 2x + \frac{1}{\pi}\sin 3x - \cdots$

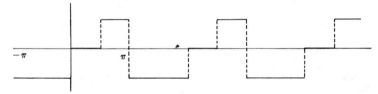

9. $f(x) = \frac{\pi}{2} - \frac{4}{\pi}\cos x - \frac{4}{9\pi}\cos 3x - \cdots$

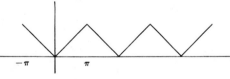

11. $f(x) = \frac{5}{2} - \frac{10}{\pi}(\sin\frac{\pi x}{3} + \frac{1}{3}\sin \pi x - \cdots)$

13. $f(t) = \frac{2}{\pi} - \frac{4}{3\pi}\cos 2t - \frac{4}{15\pi}\cos 4t - \cdots$

15. $f(t) = 2 + \frac{8}{\pi}(\cos\frac{\pi}{2}t - \frac{1}{3}\cos\frac{3\pi}{2}t + \cdots + \sin\frac{\pi}{2}t + \sin \pi t + \frac{1}{3}\sin\frac{3\pi}{2}t + \cdots)$

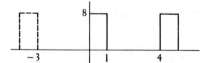

Review Exercises for Chapter 28, page 907

1. $\frac{1}{2} - \frac{1}{4}x + \frac{1}{48}x^3 - \cdots$ **3.** $2x^2 - \frac{4}{3}x^6 + \frac{4}{15}x^{10} - \cdots$ **5.** $1 + \frac{1}{3}x - \frac{1}{9}x^2 + \cdots$ **7.** $x + \frac{1}{6}x^3 + \frac{3}{40}x^5 + \cdots$

9. 0.82   **11.** 1.09   **13.** 1.0344   **15.** -0.2015 **17.** 0.952 99   **19.** 12.1655   **21.** 0.259

23. $\frac{1}{2} - \frac{1}{2}\sqrt{3}(x - \frac{1}{3}\pi) - \frac{1}{4}(x - \frac{1}{3}\pi)^2 + \cdots$

25. $f(x) = \frac{1}{2} + \frac{2}{\pi}(\cos x - \frac{1}{3}\cos 3x + \cdots)$

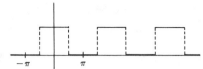

27. $f(x) = \frac{4}{\pi}(\sin \frac{\pi x}{2} - \frac{1}{2}\sin \pi x + \frac{1}{3}\sin \frac{3\pi x}{2} - \cdots)$

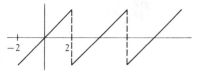

29. $(x + h) - \frac{(x+h)^3}{3!} + \cdots - (x - h) + \frac{(x-h)^3}{3!} - \cdots = 2h - \frac{2hx^2}{2!} + \cdots = 2h\left(1 - \frac{x^2}{2!} + \cdots\right)$

31. 256

33.

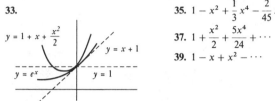

$y = 1 + x + \frac{x^2}{2}$

$y = x + 1$

$y = e^x$

$y = 1$

35. $1 - x^2 + \frac{1}{3}x^4 - \frac{2}{45}x^6 + \cdots$

37. $1 + \frac{x^2}{2} + \frac{5x^4}{24} + \cdots$

39. $1 - x + x^2 - \cdots$

41. 0.002 496 88   **43.** $N = N_0\left(1 - \lambda t + \frac{\lambda^2 t^2}{2} - \frac{\lambda^3 t^3}{6} + \cdots\right)$

45. $N_0[1 + e^{-k/T} + (e^{-k/T})^2 + \cdots] = N_0(1 + e^{-k/T} + e^{-2k/T} + \cdots)$

47. $f(t) = \frac{1}{2\pi} + \frac{1}{\pi}(\frac{1}{2}\cos t - \frac{1}{3}\cos 2t + \cdots) + \frac{1}{4}\sin t + \frac{2}{3\pi}\sin 2t + \cdots$

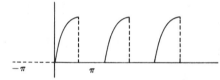

Exercises 29-1, page 912

1. Particular solution   **3.** General solution

(The following "answers" are the unsimplified expressions obtained by substituting functions and derivatives.)

5. $1 = 1$ **7.** $e^x - (e^x - 1) = 1$ **9.** $x(2cx) = 2(cx^2)$ **11.** $(-2ce^{-2x} + 1) + 2(ce^{-2x} + x - \frac{1}{2}) = 2x$

13. $(-12\cos 2x) + 4(3\cos 2x) = 0$

15. $[e^{2x}(1 + 4c_1 + 4c_2 + 4x + 4c_2 x + 2x^2)] - 4[e^{2x}(2c_1 + c_2 + x + 2c_2 x + x^2)] + 4[e^{2x}(c_1 + c_2 x + x^2/2)] = e^{2x}$

17. $x^2\left[-\frac{c^2}{(x-c)^2}\right] + \left[\frac{cx}{(x-c)}\right]^2 = 0$ **19.** $x\left(-\frac{c_1}{x^2}\right) + \frac{c_1}{x} = 0$

21. $(\cos x - \sin x + e^{-x}) + (\sin x + \cos x - e^{-x}) = 2\cos x$

23. $(e^{-x} + \frac{12}{5}\cos 2x + \frac{24}{5}\sin 2x) + (-e^{-x} + \frac{6}{5}\sin 2x - \frac{12}{5}\cos 2x) = 6\sin 2x$

25. $\cos x\left[\frac{(\sec x + \tan x) - (x + c)(\sec x \tan x + \sec^2 x)}{(\sec x + \tan x)^2}\right] + \sin x = 1 - \frac{x + c}{\sec x + \tan x}$ **27.** $c^2 + cx = cx + c^2$

Exercises 29-2, page 917

1. $y = c - x^2$ **3.** $x - \frac{1}{y} = c$ **5.** $\ln(x^3 + 5) + 3y = c$ **7.** $4\sqrt{1 - y} = e^{-x^2} + c$ **9.** $e^x - e^{-y} = c$

11. $\ln(y + 4) = x + c$ **13.** $y(1 + \ln x)^2 + cy + 2 = 0$ **15.** $\tan^2 x + 2\ln y = c$

17. $x^2 + 1 + x\ln y + cx = 0$ **19.** $y^2 + 4\arcsin x = c$ **21.** $y = c - (\ln x)^2$ **23.** $\ln(e^x + 1) - \frac{1}{y} = c$

25. $3\ln y + x^3 = 0$ **27.** $\frac{1}{3}y^3 + y = \frac{1}{2}\ln^2 x$ **29.** $2\ln(1 - y) = 1 - 2\sin x$ **31.** $e^{2x} - \frac{2}{y} = 2(e^x - 1)$

Exercises 29-3, page 919

1. $2xy + x^2 = c$ **3.** $x^3 - 2y = cx - 4$ **5.** $x^2y - y = cx$ **7.** $(xy)^4 = 12\ln y + c$ **9.** $2\sqrt{x^2 + y^2} = x + c$

11. $y = c - \frac{1}{2}\ln\sin(x^2 + y^2)$ **13.** $\ln(y^2 - x^2) + 2x = c$ **15.** $5xy^2 + y^3 = c$ **17.** $2xy + x^3 = 5$

19. $2x = 2xy^2 - 15y$

120

1. $y = e^{-x}(x + c)$ **3.** $y = -\frac{1}{2}e^{-4x} + ce^{-2x}$ **5.** $y = -2 + ce^{2x}$ **7.** $y = x(3 \ln x + c)$ **9.** $y = \frac{8}{7}x^3 + \frac{c}{\sqrt{x}}$

11. $y = -\cot x + c \csc x$ **13.** $y = (x + c) \csc x$ **15.** $y = 3 + ce^{-x}$ **17.** $2y = e^{4x}(x^2 + c)$
19. $y = \frac{1}{4} + ce^{-x^4}$ **21.** $3y = x^4 - 6x^2 - 3 + cx$ **23.** $xy = (x^3 + c)e^{3x}$ **25.** $y = e^{-x}$
27. $y = \frac{4}{3}\sin x - \csc^2 x$ **29.** $y = e^{\sqrt{x}} + (3e - e^2)e^{-\sqrt{x}}$ **31.** $y = x^2(3 + \sin x)$

1. $y^2 = 2x^2 + 1$ **3.** $y = 2e^x - x - 1$ **5.** $y^2 = c - 2x$ **7.** $y^2 = c - 2 \sin x$ **9.** $N = N_0(0.5)^{t/40}$, 35.4%

11. 4130 years   **13.** $S = a + \frac{c}{r^2}$ **15.** 1.7×10^5 **17.** 11.8 min   **19.** \$1083.29

21. $\lim_{t \to \infty} \frac{E}{R}(1 - e^{-Rt/L}) = \frac{E}{R}$ **23.** $i = \frac{E}{R^2 + \omega^2 L^2}(R \sin \omega t - \omega L \cos \omega t + \omega L e^{-Rt/L})$ **25.** $q = q_0 e^{-t/RC}$
27. 1.47 kg   **29.** $v = 9.8(1 - e^{-t})$, 9.8   **31.** 4.00 m/s   **33.** $x = 3t^2 - t^3$, $y = 6t^2 - 2t^3 - 9t^4 + 6t^5 - t^6$
35. $p = 100(0.8)^{h/2000}$ **37.** \$1260   **39.** $x = 0.20 + 0.15e^{-0.20t}$

1. $y = c_1 e^{3x} + c_2 e^{-2x}$ **3.** $y = c_1 e^{-x} + c_2 e^{-x/3}$ **5.** $y = c_1 + c_2 e^{3x}$ **7.** $y = c_1 e^{6x} + c_2 e^{2x/3}$
9. $y = c_1 e^{x/3} + c_2 e^{-3x}$ **11.** $y = c_1 e^{x/3} + c_2 e^{-x}$ **13.** $y = e^x(c_1 e^{x\sqrt{2}/2} + c_2 e^{-x\sqrt{2}/2})$
15. $y = e^{3x/8}(c_1 e^{x\sqrt{41}/8} + c_2 e^{-x\sqrt{41}/8})$ **17.** $y = e^{3x/2}(c_1 e^{x\sqrt{13}/2} + c_2 e^{-x\sqrt{13}/2})$
19. $y = e^{-x/2}(c_1 e^{x\sqrt{33}/2} + c_2 e^{-x\sqrt{33}/2})$ **21.** $y = \frac{1}{5}(3e^{7x} + 7e^{-3x})$ **23.** $y = \frac{e^3}{e^7 - 1}(e^{4x} - e^{-3x})$
25. $y = c_1 + c_2 e^{-x} + c_3 e^{3x}$ **27.** $y = c_1 e^x + c_2 e^{2x} + c_3 e^{3x}$

1. $y = (c_1 + c_2 x)e^x$ **3.** $y = (c_1 + c_2 x)e^{-6x}$ **5.** $y = c_1 \sin 3x + c_2 \cos 3x$
7. $y = e^{-x/2}(c_1 \sin \frac{1}{2}\sqrt{7}x + c_2 \cos \frac{1}{2}\sqrt{7}x)$ **9.** $y = c_1 + c_2 x$ **11.** $y = c_1 \sin \frac{1}{2}x + c_2 \cos \frac{1}{2}x$
13. $y = (c_1 + c_2 x)e^{3x/4}$ **15.** $y = c_1 \sin \frac{1}{5}\sqrt{2}x + c_2 \cos \frac{1}{5}\sqrt{2}x$ **17.** $y = e^x(c_1 \cos \frac{1}{2}\sqrt{6}x + c_2 \sin \frac{1}{2}\sqrt{6}x)$
19. $y = (c_1 + c_2 x)e^{4x/5}$ **21.** $y = e^{3x/4}(c_1 e^{x\sqrt{17}/4} + c_2 e^{-x\sqrt{17}/4})$ **23.** $y = c_1 e^{x(-6 + \sqrt{42})/3} + c_2 e^{x(-6 - \sqrt{42})/3}$
25. $y = e^{(\pi/6 - 1 - x)} \sin 3x$ **27.** $y = (4 - 14x)e^{4x}$ **29.** $(D^2 - 9)y = 0$ **31.** $(D^2 + 9)y = 0$

1. $y = c_1 e^{2x} + c_2 e^{-x} - 2$ **3.** $y = c_1 \sin \frac{1}{2}x + c_2 \cos \frac{1}{2}x + x^2 - 8$ **5.** $y = c_1 e^{-x} + c_2 e^{-3x} + \frac{1}{8}e^x + \frac{2}{3}$
7. $y = c_1 + c_2 e^{3x} - \frac{3}{4}e^x - \frac{1}{2}xe^x$ **9.** $y = c_1 e^{x/3} + c_2 e^{-x/3} - \frac{1}{10} \sin x$
11. $y = c_1 \sin 3x + c_2 \cos 3x + x + 2 \sin 2x + \cos 2x$ **13.** $y = c_1 e^{-5x} + c_2 e^{6x} - \frac{1}{3}$
15. $y = c_1 e^{2x/3} + c_2 e^{-5x} + \frac{1}{4}e^{3x}$ **17.** $y = c_1 e^{2x} + c_2 e^{-2x} - \frac{1}{5} \sin x - \frac{2}{5} \cos x$
19. $y = c_1 \sin x + c_2 \cos x - \frac{1}{3} \sin 2x + 4$ **21.** $y = c_1 e^{-x} + c_2 e^{-4x} - \frac{7}{100}e^x + \frac{1}{10}xe^x + 1$
23. $y = (c_1 + c_2 x)e^{-3x} + \frac{1}{25}e^{2x} - e^{-2x}$ **25.** $y = \frac{1}{6}(11e^{3x} + 5e^{-2x} + e^x - 5)$
27. $y = -\frac{2}{3} \sin x + \pi \cos x + x - \frac{1}{3} \sin 2x$

1. $\theta = 0.1 \cos 3.1t$

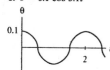

3. 10   **5.** $y = 0.100 \cos 14t$
7. $y = 0.100 \cos 14t + 0.050 \sin 2t - 0.007 \sin 14t$
9. $x = 0.150 \cos 6.26t$
11. $q = 2.23 \times 10^{-4}e^{-20t} \sin 2240t$ **13.** $q = 0.01(1 - \cos 316t)$
15. $q = e^{-10t}(c_1 \sin 99.5t + c_2 \cos 99.5t) - \frac{1}{2290}(2 \cos 200t + 15 \sin 200t)$
17. $i = 10^{-6}(2.00 \cos 1.58 \times 10^4 t + 158 \sin 1.58 \times 10^4 t - 2.00e^{-200t})$
19. $i_p = 0.528 \sin 100t - 3.52 \cos 100t$

1. $F(s) = \int_0^\infty e^{-st} dt = -\frac{1}{s}e^{-st}\Big|_0^\infty = \frac{1}{s}$ **3.** $F(s) = \int_0^\infty e^{-st} \sin at \, dt = \frac{e^{-st}(-s \sin at - a \cos at)}{s^2 + a^2}\Big|_0^\infty = \frac{a}{s^2 + a^2}$

5. $\frac{1}{s - 3}$ **7.** $\frac{6}{(s + 2)^4}$ **9.** $\frac{s - 2}{s^2 + 4}$ **11.** $\frac{3}{s} + \frac{2(s^2 - 9)}{(s^2 + 9)^2}$ **13.** $s^2 L(f) + s L(f)$
15. $(2s^2 - s + 1)L(f) - 2s + 1$ **17.** t^2 **19.** $\frac{1}{2}e^{-3t}$ **21.** $\frac{1}{2}t^2 e^{-t}$ **23.** $\frac{1}{54}(9t \sin 3t + 2 \sin 3t - 6t \cos 3t)$

Exercises 29-12, page 956

1. $y = e^{-t}$ **3.** $y = -e^{3t/2}$ **5.** $y = (1 + t)e^{-3t}$ **7.** $y = \frac{1}{2}\sin 2t$ **9.** $y = 1 - e^{-2t}$ **11.** $y = e^{2t}\cos t$
13. $y = 1 + \sin t$ **15.** $y = e^{-t}(\frac{1}{2}t^2 + 3t + 1)$ **17.** $v = 6(1 - e^{-t/2})$ **19.** $q = 1.6 \times 10^{-4}(1 - e^{-5000t})$
21. $i = 5t \sin 50t$ **23.** $y = \sin 3t - 3t \cos 3t$

Review Exercises for Chapter 29, page 957

1. $2\ln(x^2 + 1) - \dfrac{1}{2y^2} = c$ **3.** $ye^{2x} = x + c$ **5.** $y = c_1 + c_2 e^{-x/2}$ **7.** $y = (c_1 + c_2 x)e^{-x}$

9. $2x^2 + 4xy + y^4 = c$ **11.** $y = cx^3 - x^2$ **13.** $y = c(y + 2)e^{2x}$ **15.** $y = e^{-x}(c_1 \sin\sqrt{5}x + c_2 \cos\sqrt{5}x)$
17. $y = \frac{1}{2}(1 + ce^{-4x})$ **19.** $y = \frac{1}{2}(c - x^2)\csc x$ **21.** $y = c_1 e^x + c_2 e^{-3x/2} - 2$
23. $y = e^{-x/2}(c_1 e^{x\sqrt{5}/2} + c_2 e^{-x\sqrt{5}/2}) + 2e^x$ **25.** $y = c_1 e^{2x/3} + c_2 e^{4x/3} + \frac{1}{2}x + \frac{25}{8}$
27. $y = c_1 \sin 3x + c_2 \cos 3x + \frac{1}{8}\sin x$ **29.** $y^3 = 8 \sin^2 x$ **31.** $y = 2x - 1 - e^{-2x}$
33. $y = 2e^{-x/2}\sin(\frac{1}{2}\sqrt{15}x)$ **35.** $y = \frac{1}{25}[16 \sin x + 12 \cos x - 3e^{-2x}(4 + 5x)]$ **37.** $y = e^{t/4}$
39. $y = \frac{1}{2}(e^{3t} - e^t)$ **41.** $y = -4 \sin t$ **43.** $y = \frac{1}{3}t - \frac{4}{9}\sin 3t$ **45.** $r = r_0 + kt$ **47.** 3.93 m/s

49. 12.5 years   **51.** 6.5 billion   **53.** $5y^2 + x^2 = c$ **55.** $q = c_1 e^{-t/RC} + EC$
57. $y = 0.25e^{-2t}(2 \cos 4t + \sin 4t)$, underdamped   **59.** $q = e^{-6t}(0.4 \cos 8t + 0.3 \sin 8t) - 0.4 \cos 10t$
61. $i = 0$ **63.** $i = 12(1 - e^{-t/2}); i(0.3) = 1.67$ A   **65.** $q = 10^{-4}e^{-8t}(4.0 \cos 200t + 0.16 \sin 200t)$
67. $y = 0.25t \sin 8t$ **69.** $y = \dfrac{10}{3EI}[100x^3 - x^4 + xL^2(L - 100)]$

71. $\dfrac{dv}{dt} = \dfrac{dv}{dr}\dfrac{dr}{dt} = v\dfrac{dv}{dt}; \dfrac{dv}{dt} = \dfrac{k}{r^2}, k = -gR^2, \dfrac{-gR^2}{r^2} = v\dfrac{dv}{dr}, v^2 = \dfrac{2gR^2}{r} + v_0^2 - 2gR; v^2 \to 0$ as $r \to \infty$ if $v_0^2 \geq 2gR$

Exercises S-1, page 964

1. $x = 2, y = 1$ **3.** $x = \frac{17}{14}, y = \frac{19}{14}$ **5.** $x = -1, y = 4, z = 2$ **7.** $x = -2, y = -\frac{2}{3}, z = \frac{1}{3}$
9. $w = 1, x = 0, y = -2, z = 3$ **11.** Unlimited: $x = -3, y = -1, z = 1; x = 12, y = 0, z = -10$
13. Inconsistent   **15.** $x = \frac{1}{2}, y = -\frac{3}{2}$ **17.** $x = \frac{1}{6}, y = -\frac{1}{2}$ **19.** Inconsistent   **21.** 5.7 calc/s, 6.8 calc/s
23. 250 parts/h, 220 parts/h, 180 parts/h

Exercises S-2, page 969

1. Hyperbola;
$2x'y' + 25 = 0$

3. Ellipse;
$4x'^2 + 9y'^2 = 36$

5. Parabola;
$x'^2 + \sqrt{2}y' = 0$

7. Hyperbola;
$4x'^2 - y'^2 = 4$

9. Ellipse;
$x'^2 + 2y'^2 = 2$

11. Parabola;
$y'^2 - 4x' + 16 = 0$
$y''^2 = 4x''$

Exercises S-3, page 972

1. $V = \frac{1}{3}\pi r^2 h$ **3.** $R = \sqrt{F_1^2 + F_2^2 + 1.732F_1 F_2}$ **5.** 24   **7.** $2 - 3y + 4y^2$
9. $\dfrac{p^2 + pq + kp - p + 2q^2 + 4kq + 2k^2 + 5q + 5k}{p + q + k}$ **11.** 0   **13.** $x \neq 0, y \geq 0$ **15.** 1.03 A, 1.23 A
17. $A = \dfrac{pw - 2w^2}{2}$, 3850 cm^2 **19.** $L = \dfrac{(1.15 \times 10^4)r^4}{\ell^2}$

Exercises S-4, page 978

1.

3.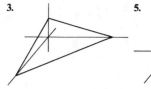

5.

7.

9.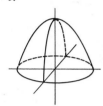

11.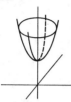

13.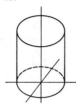

15.

17.

19.

Exercises S-5, page 983

1. $\dfrac{\partial z}{\partial x} = 5 + 8xy, \dfrac{\partial z}{\partial y} = 4x^2$ 3. $\dfrac{\partial f}{\partial x} = e^{2y}, \dfrac{\partial f}{\partial y} = 2xe^{2y}$ 5. $\dfrac{\partial \phi}{\partial r} = \dfrac{1 + 3rs}{\sqrt{1 + 2rs}}, \dfrac{\partial \phi}{\partial s} = \dfrac{r^2}{\sqrt{1 + 2rs}}$

7. $\dfrac{\partial z}{\partial x} = y \cos xy, \dfrac{\partial z}{\partial y} = x \cos xy$ 9. $\dfrac{\partial f}{\partial x} = \dfrac{12 \sin^2 2x \cos 2x}{1 - 3y}, \dfrac{\partial f}{\partial y} = \dfrac{6 \sin^3 2x}{(1 - 3y)^2}$

11. $\dfrac{\partial z}{\partial x} = \cos x - y \sin xy, \dfrac{\partial z}{\partial y} = -x \sin xy + \sin y$ 13. -8

15. $\dfrac{\partial^2 z}{\partial x^2} = -6y, \dfrac{\partial^2 z}{\partial y^2} = 12xy, \dfrac{\partial^2 z}{\partial x \partial y} = 6y^2 - 6x$ 17. $-4, -4$

19. 114 cm^2

21. $M\left(\dfrac{2mg}{(M + m)^2}\right) + m\left(\dfrac{-2Mg}{(M + m)^2}\right) = 0$

23. $3.75 \times 10^{-3}\ 1/\Omega$

Exercises S-6, page 988

1. $\frac{28}{3}$ 3. $\frac{1}{3}$ 5. 1 7. $\frac{74}{5}$ 9. $\frac{32}{3}$ 11. $\frac{28}{3}$ 13. 300 cm^3 15.

Exercises S-7, page 992

1. $\ln\left|\dfrac{(x+1)^2}{x+2}\right| + C$　**3.** $\dfrac{1}{4}\ln\left|\dfrac{x-2}{x+2}\right| + C$　**5.** $x + \ln\left|\dfrac{x}{(x+3)^4}\right| + C$　**7.** 1.057　**9.** $\ln\left|\dfrac{x^2(x-5)^3}{x+1}\right| + C$

11. 0.08495　**13.** 1.79 C　**15.** $\frac{1}{2}\ln\frac{5}{4} = 0.1116$

Exercises S-8, page 997

1. $\dfrac{3}{x-2} + 2\ln\left|\dfrac{x-2}{x}\right| + C$　**3.** $\dfrac{2}{x} + \ln\left|\dfrac{x-1}{x+1}\right| + C$　**5.** $-\frac{5}{4}$　**7.** $\frac{1}{8}\pi + \ln 3 = 1.491$

9. $-\dfrac{2}{x} + \dfrac{3}{2}\operatorname{Arctan}\dfrac{x+2}{2} + C$　**11.** $\ln(x^2+1) + \dfrac{1}{x^2+1} + C$　**13.** $2 + 4\ln\frac{2}{3} = 0.3781$　**15.** 0.9190 m

Exercises B-1, page A-9

1. MHz, 1 MHz $= 10^6$ Hz　**3.** mm, 1 mm $= 10^{-3}$ m　**5.** kilovolt, 1 kV $= 10^3$ V
7. milliampere, 1 mA $= 10^{-3}$ A　**9.** 10^5 cm　**11.** 2.0×10^{-5} Ms　**13.** 537.00°C　**15.** 2.50×10^{-4} m^2
17. 8.00×10^4 dm^3　**19.** 0.050 L　**21.** 4.50×10^3 cm/s　**23.** 1.38×10^3 m/h　**25.** 3.53×10^6 cm/min^2
27. 5.25×10^{-3} W/A　**29.** 2.2 Mg　**31.** 5.6×10^4 cm^3　**33.** 1000 g/L　**35.** 1.20×10^3 km/h
37. 2.5×10^4 mg/L　**39.** 992 m/s

Exercises B-2, page A-12

1. 8 is exact.　**3.** 3 is exact; 74.6 is approx.　**5.** 1 and 19.3 are approx.
7. 150 is approx; 76 is exact.　**9.** 3, 4　**11.** 3, 4　**13.** 3, 3　**15.** 1, 6　**17.** (a) 3.764, (b) 3.764
19. (a) 0.01, (b) 30.8　**21.** (a) Same, (b) 78.0　**23.** (a) 0.004, (b) same　**25.** (a) 4.93, (b) 4.9
27. (a) 50 900, (b) 51 000　**29.** (a) 861, (b) 860　**31.** (a) 0.305, (b) 0.31　**33.** (a) 0.950, (b) 0.95
35. (a) 31.0, (b) 31　**37.** 128.25 m, 128.35 m　**39.** 0.1640 L

Exercises B-3, page A-16

1. 51.2　**3.** 1.69　**5.** 431.4　**7.** 30.8　**9.** 62.1　**11.** 270　**13.** 160　**15.** 27 000　**17.** 5.7
19. 4.39　**21.** 10.2　**23.** 22　**25.** 2.38　**27.** 0.042　**29.** 17.62　**31.** 18.85　**33.** 196 m
35. 35 m/s　**37.** 262 144 bytes　**39.** 83.82 cm　**41.** 37°
43. Too many sig. digits; time has only two sig. digits

Exercises for Appendix C, page A-22

1. 85°　**3.** 140°　**5.** 52°　**7.** 50°　**9.** 70°　**11.** 56°　**13.** 25°　**15.** 25°　**17.** 120°　**19.** 40°
21. 5　**23.** 17　**25.** 8　**27.** 5.120　**29.** 10.44　**31.** 28.89　**33.** 21　**35.** 20　**37.** 3.3 cm, 4.0 cm
39. 16　**41.** 25 m　**43.** 10 m　**45.** 18.46 dm　**47.** 32.7 cm　**49.** 212 cm　**51.** 62.8 km　**53.** 28 cm^2
55. 13 m^2　**57.** 24 m^2　**59.** 154.8 m^2　**61.** 216 cm^3　**63.** 20 mm^3　**65.** 924 cm^3　**67.** 153 mm^3
69. 904 cm^3　**71.** 690.5 m^3　**73.** 208 cm^2　**75.** 537 m^2　**77.** 3241 m^2　**79.** 343 cm^2　**81.** 2.00 cm^3
83. 1320 m^2

Exercises for Appendix D, page A-31

(Most answers have been rounded off to four significant digits.)
1. 56.02　**3.** 4162.1　**5.** 18.65　**7.** 0.3954　**9.** 14.14　**11.** 0.5251　**13.** 13.35　**15.** 944.6
17. 0.7349　**19.** −0.7594　**21.** −1.337　**23.** 1.015　**25.** 41.35°　**27.** −1.182　**29.** 0.5862
31. 6.695　**33.** 3.508　**35.** 0.005 685　**37.** 2.053　**39.** 5.765　**41.** 4.501×10^{10}　**43.** 497.2
45. 6.648　**47.** 401.2　**49.** 8.841　**51.** 2.523　**53.** 10.08　**55.** 22.36　**57.** 20.3°　**59.** 4729
61. 3.301×10^4　**63.** 1.056　**65.** 55.5°　**67.** 3.277　**69.** 8.125　**71.** 1.000　**73.** 1.000　**75.** 12.90
77. 8.001　**79.** 8.053　**81.** 0.042 59　**83.** 0.4219　**85.** 0.7822　**87.** 2.073 646 5　**89.** 124.3　**91.** 252

Exercises for Appendix E, page A-40

1. 100 DEF FN F(X) = 2*X/(X^2 + 1)

3. Change B to C in lines 30 and 100; delete line 160
130 INPUT "SIDE C = ";SC
145 LET SB = SQR(SC^2 − SA^2)
210 PRINT "SIDE B = ";SB

5. 190 PRINT "X1 = (";−B;" + SQR(";D;"))/";2*A
195 PRINT "X2 = (";−B;" − SQR(";D;"))/";2*A

7. 125 IF SA > SB + SC OR SB > SA + SC OR SC > SA + SB THEN GOTO 225
   222 GOTO 230
   225 PRINT "NO TRIANGLE HAS THESE THREE SIDES"

9. 120 INPUT "R = ";R:INPUT "T = ";T
   140 LET T = 3.14159265*T/180
   150 PRINT R*COS(T);" + (";R*SIN(T);")J"
   160 END

11. Change line 20
    195 IF X = −D/C THEN GOTO 285
    200 LET Y = (A*X + B)/(C*X + D)
    285 PRINT "F(X) IS UNDEFINED"
    287 GOTO 140

13. 150 LET LS = TAN(X)*SIN(X)/COS(X)
    160 LET RS = 1/(COS(X))^2 − 1

15. 140 DEF FN N(X) = 2 − SQR(X + 2)
    150 DEF FN D(X) = X − 2

17. 100 DEF FN F(X) = 4*X − X^2
    10.66, 0.59; −10.07; curve is below axis for $x > 4$

19. 140 FOR N = 1 TO 20
    On most computers values will increase past the value of e, and then at some point all values become 1. The way in which numbers are calculated on a computer leads to an error when the numbers are extremely small.

SUPPLEMENTARY TOPIC

PROBABILITY

Decisions which we make about the course of action to be taken now or in the future are based on knowledge which involves at least some uncertainty as to the outcome. For example, a decision as to whether to play golf would normally be based on the probability of good weather. In the same way, knowledge of events in the past is usually incomplete, for not every fact about such events is obtainable. This can be illustrated in the statement that "from available information, the age of the Earth is estimated to be about four billion years." Also, events in recorded history often have contradictions and varied interpretations as to actual happenings.

The basic concepts of probability are widely used, at least in an intuitive way. In this section we deal with making determinations as to possible outcomes of events. We will consider certain basic types of situations in which the probability of outcomes can be determined by the nature of the event, or on what is known from past experience.

In the study of probability, a numerical value is given to the likelihood of some particular event actually happening. In determining such a value, we assume that all events are equally likely to occur, unless we have special knowledge to the contrary.

Example A
In considering the possible outcomes of the toss of a coin, we assume that the coin will land either heads or tails, and that either of these possibilities is equally likely.

When considering the drawing of a card from a deck of cards, we assume that the deck is thoroughly shuffled, and that any of the cards in the deck is as likely to be chosen as any other.

When a die is tossed, it is assumed that any of the six faces will come up on top with equal likelihood.

Example B
If a study were made of the percentages of usable and defective parts produced by a particular machine, it would normally be expected that the machine would not turn out as many defective as usable parts. Thus, by studying a random group of parts, we can count the number of defective parts produced. This number would then form a basis of the probable percentage of defective parts which the machine would produce.

Probability, as used in this chapter, is defined as follows: *If an event can turn out in any one of n equally likely ways, and s of these would be successful, then the probability of the event occurring successfully is*

$$P = \frac{s}{n} \qquad \qquad \text{(S-1)}$$

If the number of equally likely ways an event may turn out cannot be determined from theoretical considerations and N trials are made, of which S are successful, the probability of success is given by

$$P = \frac{S}{N} \qquad \qquad \text{(S-2)}$$

If the events are equally likely, as expressed in Eq. (S-1), the probability is called an *a priori* probability. This means that probabilities of possible outcomes are determined without any experimentation, and are based on a knowledge of the nature of the event. If the probability is based on past experience, as expressed in Eq. (S-2), the probability is termed **empirical probability.**

Example C

When a card is drawn from a bridge deck (the standard 52-card deck), the probability that this card is a diamond is $\frac{1}{4}$. Here we know that of the 52 cards, 13 of them are diamonds, and that the drawing of a diamond is a success. This in turn means that $n = 52$ and $s = 13$. Thus, for this case,

$$P = \frac{13}{52} = \frac{1}{4}$$

In the same way, the probability of drawing an ace is $\frac{1}{13}$, since there are 4 aces in the 52 cards. For this case, $n = 52$ and $s = 4$, which means

$$P = \frac{4}{52} = \frac{1}{13}$$

Example D

If a particular machine produced 20 defective parts from a lot of 1000, the empirical probability of a defective part being produced is $\frac{1}{50}$. In this case $N = 1000$ and $S = 20$. Thus,

$$P = \frac{20}{1000} = \frac{1}{50}$$

It should be pointed out here that the larger the number inspected and counted, the more accurate is the empirical probability. However, the number sampled must not be so large that it is impractical.

It should be noted that when we have calculated the probability of a certain event occurring, there is no assurance that it will, or will not occur as indicated by the value of the probability. For example, in Example D, we should not expect that exactly one of every fifty parts will be defective. However, we should expect that the more parts we consider, the more likely that the ratio of defective parts to total parts will be approximately $\frac{1}{50}$.

We can see from the definitions and examples that the value of a particular probability can extend from 0 (impossible) to 1 (the sure thing). A probability of $\frac{1}{2}$ expresses equal likelihood of success or failure.

There are cases in which it is more practical to compute the probability of the failure of an event, in order to calculate the probability of its success. This is based on the fact that the probability of failure is

$$F = 1 - \frac{s}{n}$$

Thus, the probability of success, in terms of the probability of failure, is

$$\boxed{P = 1 - F} \qquad \text{(S-3)}$$

129

Example E

A bag contains 3 red balls, 4 white balls, and 7 black balls. What is the probability of drawing a red or a black ball?

This can be computed directly, or it can be computed by first determining the probability of drawing a white ball, which would be the probability of failure.

Computing directly, we know that there are 10 balls which are either red or black. Thus, $s = 10$ and $n = 14$, which means that

$$P = \frac{10}{14} = \frac{5}{7}$$

Now, computing the probability of failure first, we know that there are 4 balls (the white ones) which would not be successful draws. Thus,

$$F = \frac{4}{14} = \frac{2}{7}$$

Using Eq. (S-3), we now have

$$P = 1 - \frac{2}{7} = \frac{5}{7}$$

We see that the results agree.

So far we have considered only the probability of success of a single event. The following examples illustrate how the probability of success of a combination of events is found.

Example F

What is the probability of a coin turning up heads in each of two successive tosses?

We can use the definition of probability to determine this result. On the first toss there are two possibilities. On the second toss there are also two possibilities, which means there are, in all, four possible ways in which the coin may fall in two successive tosses. These are HH, HT, TT, TH. Only one of these is successful (heads on two successive tosses). Thus, the probability is $\frac{1}{4}$. This is equivalent to multiplying the probability of success of the first toss by the probability of success of the second toss, or $(\frac{1}{2})(\frac{1}{2}) = \frac{1}{4}$. When we use this multiplication method it is not necessary to figure out all possibilities, a procedure which is often very lengthy, or even impossible from a practical point of view. This multiplication may be stated roughly as "there is a probability of $\frac{1}{2}$ (the second toss) of a probability of $\frac{1}{2}$ (the first toss) of success."

Based on the discussion and results of Example F, the probability of success of a compound event is given in terms of the probabilities of the separate events by

$$P_{1 \text{ and } 2} = P_1 P_2 \qquad\qquad (S-4)$$

Example G
Two cards are drawn from a deck of bridge cards. If the first card is not replaced before the second is drawn, what is the probability that both will be hearts?

The probability of drawing a heart on the first draw is $\frac{13}{52}$, or $\frac{1}{4}$. If this first card is a heart, there are only 12 hearts of 51 remaining cards for the second draw. Thus, the probability of success on the second draw is $\frac{12}{51}$. Multiplying these results, we have the probability of success for both, or

$$P = \left(\frac{1}{4}\right)\left(\frac{12}{51}\right) = \frac{1}{17}$$

This means there is a 1-in-17 chance of drawing two successive hearts in this manner.

Example H
In Example G determine the probability that both cards are hearts if the first card is replaced in the deck before the second card is drawn.

Again, the probability of drawing a heart on the first draw is $\frac{1}{4}$. However, since this card is replaced in the deck before the second draw is made, the probability that the second card is a heart is also $\frac{1}{4}$. Thus, the probability that both cards will be hearts is

$$P = \left(\frac{1}{4}\right)\left(\frac{1}{4}\right) = \frac{1}{16}$$

As we should expect, the probability of drawing two hearts in this way is slightly better than when the first card is not replaced.

Example I
In three tosses of a single die, what is the probability of tossing at least one 2?

Instead of calculating the various combinations, it is easier to calculate the probability of failure to get a 2, and subtract that from 1. This is due to the fact that the probability of failure is $\frac{5}{6}$ each time, and this must occur three times successively for failure of the compound event. Thus,

$$F = \left(\frac{5}{6}\right)\left(\frac{5}{6}\right)\left(\frac{5}{6}\right) = \frac{125}{216} \qquad \text{or} \qquad P = 1 - \frac{125}{216} = \frac{91}{216}$$

131

Example J
In eight tosses of a coin, what is the probability of tossing at least one head?

Again, it is easier to calculate the probability of failure to obtain the result. The probability of failure (tails) is $\frac{1}{2}$ for each toss. This must occur 8 successive times in order that a head does not appear. Thus,

$$F = \left(\frac{1}{2}\right)^8 = \frac{1}{256}$$

which means that

$$P = 1 - F = 1 - \frac{1}{256} = \frac{255}{256}$$

One misconception is often encountered when we are talking about probability: *In dealing with the probability that a single event will occur, we should remember that the occurrence of this event in the past does not alter the probability of the event occurring in the future.*

Example K
If a coin is tossed 7 times and it comes up tails each time, the probability that it will come up heads on the next toss is still $\frac{1}{2}$. On any given toss, the probability is $\frac{1}{2}$. In the previous example we showed that the probability of heads coming up at least once in 8 tosses is $\frac{255}{256}$. This, however, is a different problem from that of finding the probability of heads coming up on a particular toss of the coin.

The discussion of probability here has been restricted to certain basic cases. The probability of numerous other types of events can be determined. For example, it is possible to determine the probability of heads coming up exactly three times in five tosses of a coin. The general analysis of such cases, other than by specifying all possibilities, is beyond the scope of this discussion.

Exercises

In Exercises 1 through 8 consider a bag which contains 5 red balls, 6 white balls, and 9 black balls. What is the probability of drawing each of the following?

1. A red ball
2. A white ball
3. A white or black ball
4. A red or white ball
5. Two red balls on successive draws, if the first ball is replaced before the second draw is made
6. Two white balls on successive draws, if the first ball is replaced before the second draw is made

7. Two red balls on successive draws if the first ball is not replaced before the second draw is made

8. A red ball and then a white ball if the first ball is not replaced before the second draw is made

In Exercises 9 through 16 assume that we are tossing a single die. What is the probability of tossing the following?

9. A 4
10. A 2 or 4
11. Other than a 4
12. Other than a 2 or 4
13. Two successive 4's
14. Three successive 4's
15. At least one 4 in two successive tosses
16. At least one 4 in four successive tosses

In Exercises 17 through 24 assume that we are tossing two dice. An analysis of the possibilities shows that there are 36 different ways in which the dice may fall. What is the probability of tossing the following totals on the dice?

17. 2
18. 3
19. 7
20. 10
21. 7 or 11
22. 10, 11, or 12
23. 7 on two successive tosses
24. At least one 7 in two successive tosses

In Exercises 25 through 28 use the following information. An insurance company, in compiling mortality tables, found that of 10 000 10-year olds, 6 900 lived to be 60, and 4 700 lived to be 70.

25. What is the probability of a 10-year-old living to the age of 70?
26. What is the probability of a 60-year-old living to the age of 70?
27. What is the probability of two 10-year-old people living to the age of 60?
28. What is the probability of two 60-year-old people living to the age of 70?

In Exercises 29 through 32 use the following information:
The assembly of a certain product is done in three stages by machines A, B, and C. If any machine breaks down, the assembly cannot be done. The probabilities that the machines will not break down in a year are $\frac{4}{5}$, $\frac{9}{10}$, and $\frac{19}{20}$, respectively. Find the probabilities that production will cease at some point in a year due to breakdowns in the indicated machines.

29. A or B
30. B or C
31. A or C
32. Any of the machines

In Exercises 33 through 40 solve the given problems in probability.

33. What is the probability of drawing an ace on each of two successive draws from a standard bridge deck of cards, if the first card is not replaced before drawing the second card?

34. What is the probability of drawing at least one ace in two draws, if the first card is not replaced before drawing the second card?

35. For the machine in Example D, how many defective parts should we expect to find in 300 total parts?

133

36. A certain college, which accepts students only from the upper quarter of the high-school graduating class, finds that 12% of its students attain an average of 3.0 or better (based on a highest attainable score of 4.0). If all the graduates from a particular high school were to apply to this college, what is the probability of a particular one of them attaining an average of 3.0 or better in his classes at the college?

37. One of the first three trials in a series of complicated scientific experiments was unsuccessful due to a failure in a piece of equipment. Based on these three trials, what is the probability of success of the next two successive trials?

38. In Exercise 37, if the fourth trial is successful, what is the probability at this point of the fifth trial being successful?

39. In a random test of newly manufactured transistors, if there are two defective transistors in a particular group of 20, and two of the 20 are tested, what is the probability that at least one of those tested will be defective?

40. How many of the transistors of Exercise 39 would have to be tested in order for there to be a 50% chance of testing one of the defective ones?

ANSWERS TO EXERCISES

1. $\frac{1}{4}$ 2. $\frac{3}{10}$ 3. $\frac{3}{4}$ 4. $\frac{11}{20}$ 5. $\frac{1}{16}$ 6. $\frac{9}{100}$ 7. $\frac{1}{19}$ 8. $\frac{3}{38}$

9. $\frac{1}{6}$ 10. $\frac{1}{3}$ 11. $\frac{5}{6}$ 12. $\frac{2}{3}$ 13. $\frac{1}{36}$ 14. $\frac{1}{216}$ 15. $\frac{11}{36}$

16. $\frac{671}{1296}$ 17. $\frac{1}{36}$ 18. $\frac{1}{18}$ 19. $\frac{1}{6}$ 20. $\frac{1}{12}$ 21. $\frac{2}{9}$ 22. $\frac{1}{6}$

23. $\frac{1}{36}$ 24. $\frac{11}{36}$ 25. 0.470   26. 0.681   27. 0.476   28. 0.464

29. $\frac{7}{25}$ 30. $\frac{29}{200}$ 31. $\frac{6}{25}$ 32. $\frac{79}{250}$ 33. $\frac{1}{221}$ 34. $\frac{33}{221}$

35. 6   36. 0.03   37. $\frac{4}{9}$ 38. $\frac{3}{4}$ 39. 0.195   40. 6

Instructor's Guide

A Manual

To Accompany

TECHDISK

by

Robert Seaver

and

William N. Thomas Jr.

A Manual To Accompany TECHDISK

Special Instructions for Certain TECHDISK Programs

Several of the programs on TECHDISK have results that are randomly generated. For these programs, the user is asked to enter a beginning number which is used to initialize the random number generator. Each time the same positive value is entered for this initial number, the same results will be obtained. However, each time a negative value is entered for this initial number, different results will be obtained. Thus, when positive numbers are entered, the output of these programs and the conclusion reached by a student may be checked by the instructor.

Some of the conclusions which the student might be expected to reach are provided as 'answers' in the program itself. To access these 'answers', the user needs to enter the symbol @ while the program menu is on the screen. A request will then appear at the top of the screen for a password. Upon entering the correct password, the @ symbol will appear in the upper left corner of the screen. Then, when the program is run, the answer will appear on the screen along with the information the student normally sees. A choice will also appear on the menu allowing the user to run the program with other random numbers.

The following programs have this alternative:

SECTION/OPTION	PROGRAM	PASSWORD	ANSWER
Sect. 1/Opt. 2 (Use in Exercise 1.2)	Use function machine with random functions	rf	Equation of function
Sect. 3/Opt. 4 (Use in Exercise 3.4)	Work with signs of random functions	sf	Equation of function
Sect. 4/Opt. 4 (Use in Exercise 4.3)	Simulate tossing of dice	drl	Mean and standard deviation
Sect. 4/Opt. 6 (Use in Exercise 4.4)	Get random points to fit a straight line	rp	Equation of line
Sect. 5/Opt. 4 (Use in Exercise 5.3)	For a random function, get signs of derivatives	fds	Equation of function

Exercise 1.1

2. (-2,12), (-1,6), (0,2),
   (1,0), (2,0), (3,2), (4,6)

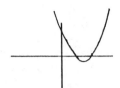

4. (5,2), (4,1.73), (3,1.41),
   (2,1), (1,0), (0,undef),
   (-2,undef)

6. (-2,1.73), (-1,0),
   (-0.9,undef), (0,undef)
   (0.5,undef), (1,0),
   (2,1.73), (3,2.83)

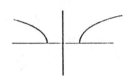

8. (5,1.33), (4,2), (3,4),
   (2.5,8), (2.1,40.0),
   (2,undef), (1.9,-40.0),
   (1.5,-8), (1,-4), (0,-2)

10. (-4,0.08), (-3,0.2),
    (-2.5,0.44), (-2,undef),
    (-1.5,-0.57), (-1,-0.33),
    (0,-0.25), (1.5, -0.57),
    (2, undef), (2.5, 0.44),
    (3, 0.20), (4, 0.08)

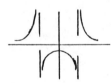

12. f(0) = -5, f(1) = -4,
    f(2) = -1, f(2.2) = -.16,
    f(3) = 4, f(4) = 11,
    f(-2) = -1, f(-2) = 4
    Domain: All real numbers.

14. f(4) = 2.45, f(2) = 2,
    f(1) = 1.73, f(0) = 1.41,
    f(-1) = 1, f(-2) = 0,
    f(-3) = undef, f(-4) = undef
    Domain: reals >= -2

16. f(-2) = undef, f(-1.6) = undef, f(-1.5) = .71, f(-1) = 1.73,
    f(0) = 2.24, f(1) = 1.73, f(1.5) = .71, f(1.6) = undef,
    f(2) = undef, Domain: -1.58 <= reals <= 1.58

18. f(-4) = .57, f(-3.1) = 6.56, f(-3) = undef, f(-2.9) = -6.78,
    f(-2) = -.8, f(0) = -.44, f(2) = -.8, f(2.9) = -6.78,
    f(3) = undef, f(4) = .57, Domain: all reals except -3 and 3.

20. (-3,-14), (-2,-11), (-1,-8),
    (0,-5), (1,-2), (2,1), (3,4),
    (4,7)

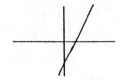

22. (-3,7), (-2,4), (-1,1)
    (0,-2), (1,-5), (2,-8),
    (3,-11)

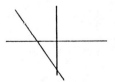

24. (-2,10), (-1,4), (0,0),
    (1,-2), (1.5,-2.25), (2,-2),
    (3,0), (4,4)

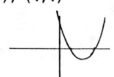

26. (-6,7), (-5,0), (-4,-5),
    (-3,-8), (-2,-9), (-1,-8),
    (0,-5), (1,0), (2,7)

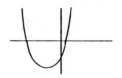

28. (-3, -32), (-2,-13), (-1,-6)
    (0,-5), (1,-4), (1.5,-1.63),
    (2,3), (3,22)

30. (-2,-150), (-1,-28),
    (-0.5,0), (0,12), (1,6),
    (1.5, -3), (2,-10), (3,0),
    (4,72)

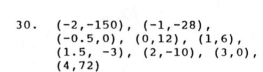

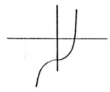

32. (-3,4.58), (-2,2.45),
    (-1.5,0.87), (-1.4,undef),
    (-1,undef), (0,undef),
    (1,undef), (1.4,undef),
    (1.5,0.87), (2,2.45),
    (3,4.58)

34. (-2,0.8), (-1,1), (0,1.33)
    (1,2), (2,4), (2.5,8),
    (3,undef), (3.5,-8),
    (4,-4), (5,-2)

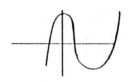

36. (-3,-0.33), (-2,-1), (-1.5,-3.33),
    (-1.23,-193.8), (-1.22,215.5),
    (-1,5), (0,1.67), (1,5),
    (1.22, 215.5), (1.23,-193.8),
    (2,-1,), (3,-0.33)

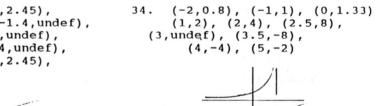

38. (a) f(2) = 6, f(3.5) = 3.43
    (b) at 8 ohms, 1.5 amps
        at 28 ohms, 0.43 amps
    (c) current becomes large

40. (a) f(3.50) = 32.01
        f(5.00) = 30.82
    (b) 29 tapes, 26, tapes
    (c) $24.00

Exercise 1.2

	(a)	(b)	(c)	(d)
2.	$4x + 1$	$3x^2 - 9$	$7x^2 - 9x$	$(-5x-2)/(x+5)$
4.	$2x + 6$	$3x^2 - 25$	$x^3 - 9x$	$(2x-5)/(x+3)$
6.	$4x - 8$	$-x^2 + 25$	$-9x^3 - 9$	$(-2x-7)/(x-1)$
8.	$-5x - 1$	$2x^2 + 1$	$-x^3 + 25x$	$7/(x-2)$

Exercise 3.1

2. 2.442, 0.279 ± 1.249j
6. -0.500, 1.353,
    -0.177 ± 1.203j
10. -0.268, -3.732, -3.333, 2 ± j
12. -0.250, 2.236, -2.236
16. ± 2.236, 1 ± j
18. 0.268, 2.726, -1.497 ± 0.702j
20. -1.000, 4.000
24. -4.000, 2.333, -0.333
28. 13.400, 38.914

4. -0.600, 2.828, -2.828
8. 0.544, 1.377, 2.392
    -1.823 ± 0.627j

14. -0.176, 2.176,
    2.902, -0.902

22. 0.606, -1.074
26. 2.236, -2.236
30. 8.074, -1.613, -0.461

Exercise 3.3

2. x = 5.769, y = -5.615
6. x = 9.667, y = 11.600,
    z = 2.133

4. x = -0.243, y = 0.824
8. x = 8.545, y = 9.739
    z = -14.551

10.
$$\begin{bmatrix} -0.171 & 0.098 \\ 0.122 & 1.173 \end{bmatrix}$$

12.
$$\begin{bmatrix} -0.117 & 0.039 & 0.143 \\ 0.156 & -0.052 & 0.143 \\ 0.299 & 0.234 & -0.143 \end{bmatrix}$$

14.
$$\begin{bmatrix} -0.193 & -0.092 & 0.135 \\ 0.252 & 0.034 & 0.042 \\ 0.202 & 0.227 & 0.034 \end{bmatrix}$$

16. x = -14.5, y = 7.33
18. x = -0.5, y = 4.0, z = -5.0
20. x = 0.223, y = 0.0015,
    z = -0.052

22. The forces are: 20.2, 14.5, and 10.7

Exercise 3.4

(a)	(b)	(c)
2. f(x)>0: x<-2 f(x)<0: x>-2	f(x)>0: x<1.75 f(x)<0: x>1.75	f(x)>0: -1<x<2 f(x)<0: x<-1, x>2

4. f(x)>0: x>4
    f(x)<0: x<4

f(x)>0: -2<x<1
f(x)<0: x<-2, x>1

f(x)>0: none
f(x)<0: x≠-1.5

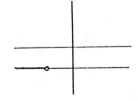

6. f(x)>0: x>0   f(x)>0: -1.7<x<-1, x>2   f(x)>0: none
    f(x)<0: x<0   f(x)<0: x<-1.7, -1<x<2   f(x)<0: x≠2

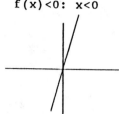

8. f(x)>0: -7<x<6   f(x)>0: x<-2,   f(x)>0: -3<x<-2,
                                       x>-0.5   x>0.5
    f(x)<0: x<-7,   f(x)<0: -2<x<-0.5   f(x)<0: x<-3,
            x>6   -2<x<0.5

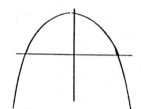

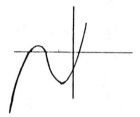

10. f(x)>0: x<-0.5,   f(x)>0: x<-1,   f(x)>0: 2<x<3
            x>1   -0.5<x<0.5
    f(x)<0: -0.5<x<1   f(x)<0: -1<x<-0.5,   f(x)<0: x<2, x>3
                                     x>0.5   asym: x = 3

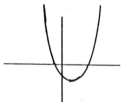

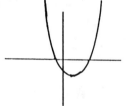

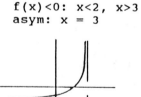

12. zero: 0.75   zero: -1.67   zeros: 0.5, 1.5

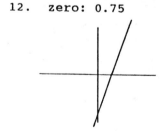

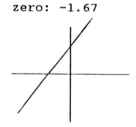

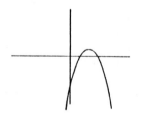

14. zeros: 1.25, 5   zeros: -0.67, 0, 1.33   zero: -4, asym: x = -5

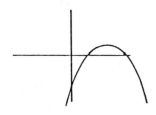 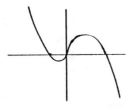 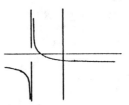

16. zeros: 3, 6   zero: 1   zeros: -1.67, -1

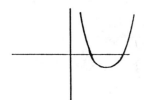 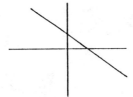 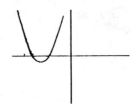

18. zeros: 0.33, 1.67   zeros: -2.67, 1, 2.5   zero: -1.33
 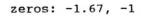 asym: x = -1.5

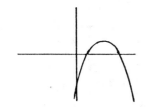 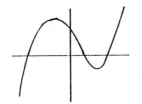 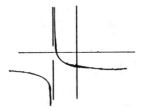

20. zeros: -6, -1.33   zeros: 0.33, 2, 3   zero: 0, axym: x = 0.5

 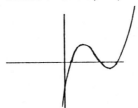 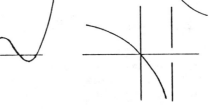

22. zeros: -6, -2   zeros: -6, -0.67, 5   zero: 0, axym: x = 0.5

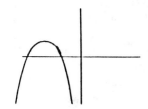 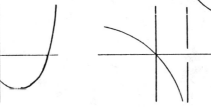

Exercise 4.1

2. Centers on y-axis; Moving k, move circles vertically; C(0,3)

4. (a) C(0,0), r = 1
   (b) C(2,2), r = 1
   (c) C(-2,2), r = 1
   (d) C(2,-1), r = 2

6. As b decreases, the ellipse tends to flatten, approaching a line segment.

8. As b increases, the hyperbola tends to flatten. As b becomes very large, graph approaches 2 horizontal lines. As b becomes small tends toward a line with a segment missing.

10. As a becomes smaller, gets closer to origin. The value a is the distance from the center to a vertex.

12. Open to right. p smaller makes graph narrower. As p becomes very small, tends toward a half line.

14. Opens to left. As B becomes large, graph flattens. As B becomes very large, tends toward a line.

16. Change in k moves graph vertically. In (d) graph opens to left.

18. Change in h moves graph horizontally. Change in a, changes width of graph. In (d), graph opens to left.

Exercise 4.2

2. (a) circle, C(3π/2,-1),r=1
   (b) max = 2, min = 0
   (c) (0,0), (π/2,2),
       (π,0), (3π/2,2)

4. (a) rose, 3 leaves
   (b) max = 1, min = 0
   (c) (0,0), (π/2,-1)
       (π,0), (3π/2,1)

6. (a) limacon
   (b) max = 3, min = 1
   (c) (0,3), (π/2,2),
       (π,1), (3π/2,2)

8. (a) limacon
   (b) max = 5, min = 0
   (c) (0,5), (π/2,2)
       (π,-1, (3π/2,2)

10, (a) hyperbola, V(π/2,1)
    V(π/2,1/3), F(0,0), F(π/2,4/3)
    (b) max = ∞, min = 1/3
    (c) (0,1), (π/2,1/3)
        (π,1), (3π/2, -1)

12. (a) spiral
    (b) max = ∞, min = 0
    (c) (0,0), (π/2, 1/4)
        (π,1), (3π/2, 2.2)

14. rotation of graph

16. 3-leaf rose, 5-leaf rose
    b odd: leaves = b
    b even: leaves = 2b

18. (a) circle, max. r = 2
    (b) rose, 16 leaves,
        mas. r = 2

    (c) rose, 8 leaves,
        max. r = 2

20. All give limacons.
    |a|>|b|: doesn't
             contain pole
    |a|=|b|: cardioid

    |a|<|b|: has inner loop

22. (0,0), (π/4,1), (3π/4,1) [or (7π/4,-1)]

24. (7π/6,3/2) [or (π/6,-3/2)], (11π/6,3/2) [or (5π/6,-3/2)]

26. (0,0)

Exercise 4.3

2. 8.80, 1.94   4. 8.31, 2.52   6. 81.9, 14.2
8. 80.8, 8.9   10. Group in #6 has higher mean,
12. 16.37, 0.27   group in #8 are closer.
14. mean = 2024, std.dev. = 63. 68.2%, 15.9%
16. mean = 0.1216, std.dev. = 0.0043, 86.6%
18. (a) 3.44, 1.67   (b) 68.1%   (c) 100%,
20. (a) 6.98, 2.39   (b) 67.6%   (c) 94.5%
22. (a) 10.47, 2.96   (b) 67.9%   (c) 96.8%

Exercise 4.4

2. $x = 0.838t + 10.22$ 4. $y = -24.3x + 1440.$
6. $y = -2.97x + 62.0$ 8. $y = -0.614x + 16.9$
12. $R = 0.074T + 11.6$ 14. $R = 0.00373/d^2 + 0.00159$
    (a) 11.6 ohms   (b) 13.1 ohms
16. $\log R = 0.029398t - 1.1451$
    (c) 16 cents, 2 cents   (e) 62 cents, $2.41

Exercise 5.1

[Note: dne = does not exist]
Gives: left hand limit, right hand limit, limit.

2. 11, 11, 11; -5, -5, -5   4. 6, 6, 6; 0, 0, 0
6. -1, -1, -1; -231, -231, -231   8. -5, -5, -5; -4, -4, -4
10. 2357, 2357, 2357; -21211, -21211, -21211
12. 0, dne, dne; 2.45, 2.45, 2.45
14. 1.0, 1.0, 1.0; dne, dne, dne
16. 0.55, 0.55, 0.55; 11.3, 11.3, 11.3
18. 1.44, 1.44, 1.44; dne, dne, dne
20. 0.0, 0.0, 0.0; 13.0, 13.0, 13.0
22. undefined, -5.0, -5.0, -5.0; undefined, dne, dne, dne
24. undefined, dne, dne, dne; undefined, 0.50, 0.50, 0.50
26. undefined, dne, dne, dne; undefined, 2.17, 2.17, 2.17
28. 11.25, 11.25, 11.25, 11.25; undefined, 8.0, 8.0, 8.0
30. -1010, -1010, -1010, -1010; undefined, dne, dne, dne
32. 15.0, 15.0, 15.0, 15.0; undefined, dne, 0.0, dne
34. 17.8, 17.8, 17.8, 17.8; undefined, dne, 0.0, dne

Exercise 5.2

2. 1.0, 1.0   4. 0.0, 8.0   6. 0.0, 0.0
8. 0.67, 10.6   10. 175, 700   12. 1.0, 1.0
14. 160., 0.016   16. 0.022, dne   18. 1.50, 2.67
20. (a) 0.0, 4.0, -4.0 (b) 0.0, 4.0, -4.0 (c) 0.0, 4.0, -4.0
22. 9.0, 0.0009   24. 50.0, 0.0   26. -77.6, -76.7

Exercise 5.3

2. (a) zero: 1/2   (b) zeros: -2, -7/3   (c) zeros: -3/2,
                                                             -1/2, 5/3

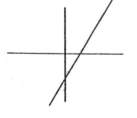

4. (a) zero: 3   (b) zeros: -2/3, 4   (c) zero: none
                                                         asym: x = -1/3

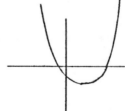
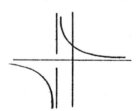

144

6. (a) zeros: 1/2, 1   (b) zeros: -3, 1, 1   (c) zero: -3/2
                                               asym: x = -1/2

 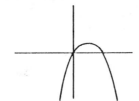 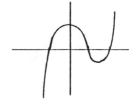

8. (a) zeros: 2/3, 1   (b) zeros: 0, 1   (c) zeros: -1, 1, 5/2

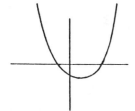 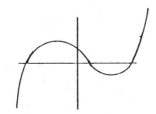 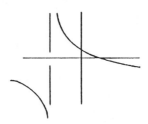

10. (a) zeros: -1/2, 3/2   (b) zeros: -5, 1, 5   (c) zero: 5/3
                                                  asym: x = -2

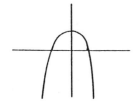 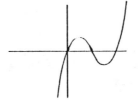 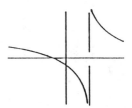

12. (a) zeros: -1/2, 1/2   (b) zeros: 0, 2, 2/3   (c) zero: -1/2
                                                   asym: x = 3/2

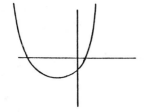 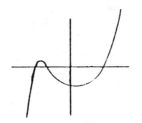 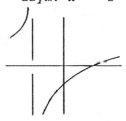

14. (a) zeros: 1/2, -5   (b) zeros: -1, -3/4, 1   (c) zero: 2
                                                   asym: x = -2

16.   (a) f(x) = 0   (b) zeros: -6, -3   (c) zero: 5/4
                                                           asym: x = 8/3

 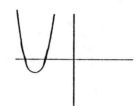

18.   (a) zeros: 3, -4   (b) zeros: -5/3, 1/4   (c) zero: 7/3
                                                        asym: x = -1

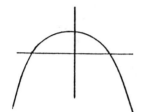 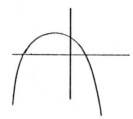 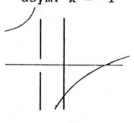

20.   (a) zeros: -8/3, -3/4   (b) zeros: 1, -3/2,   (c) zero: -3/2
                                         -3/2   asym: x = 3

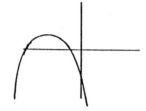 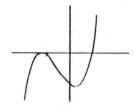 

Exercise 5.4

2.	32	4.	4.5	6.	1.377	8.	2.459
10.	0.213	12.	-3.75	14.	discontinuous		
16.	1.320	18.	1.646	20.	(a) 0 (b) 2 (c) 0 (d) 2; 0		
22.	11.8	24.	1265				

Baseball

Trace Taylor

This is a baseball.

This is a baseball field.

These are baseball players.

pitcher

The players throw
the baseball.

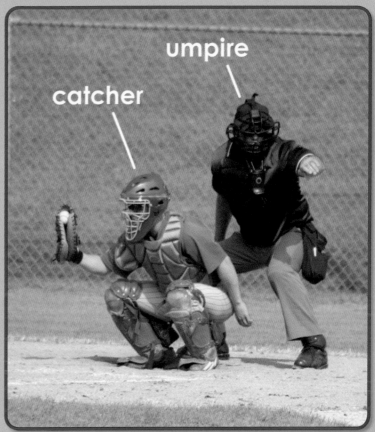

umpire

catcher

The players catch
the baseball.

batter

The players hit the ball. They hit it with a bat.

Some players don't hit the ball.

bat

runner

Players who hit the ball have to run.

This is a base.

The player has to get to the base.

Some players slide in the dirt to get there.

Some players jump to get there.

This man will see if the play is good or bad.

strike 1

This man says you didn't hit the ball.

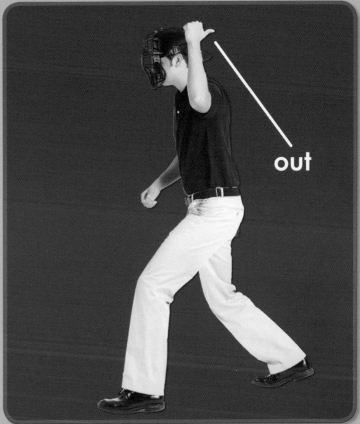

out

This man says you are out of here.

13

The Baseball

On game day, the umpire receives a box of new baseballs from the home team. Each box is sealed and certified by the league president. Before the balls are used in the game, each ball is rubbed with "magic mud" to remove its shine and make it less slippery. The mud was discovered by White Sox infielder Lena Blackburne in the 1930s. It was found along the Delaware River in New Jersey and has been used throughout the major and minor leagues ever since. Other methods have been tried to de-slick baseballs, but nothing works as well as Blackburne Baseball Rubbing Mud.

When the mud rubbing is done, the balls are taken out to the field, where each one is used for only 5 to 7 pitches. When a ball gets too dirty or scruffy, the umpire takes it out of the game and puts in another new ball. As many as 100 or more baseballs are used in one major league game.

Baseball Diamond

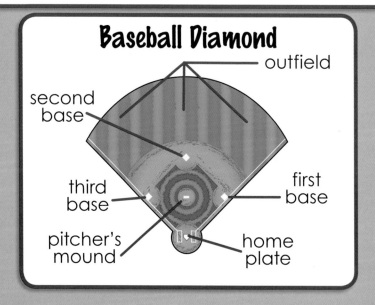

outfield

second base

third base

first base

pitcher's mound

home plate

Power Words
How many can you read?

a	has	it	see	this
are	have	jump	some	to
didn't	here	of	the	who
don't	if	or	there	will
get	in	out	these	with
good	is	says	they	you

Practice With National Reading Standards

1. What was this book about? How do you know? (CCSS 2)

2. What position would you want to play in baseball? What would you have to do in that position? What in the pictures or words supports your answer? (CCSS 3)

3. Why would a player need to slide or jump to get on base? What in the pictures or words supports your answer? (CCSS 1,7)

For more information about the National Reading Standards, please visit
www.americanreadingathome.com/common-core-standards